List of Tables

List of Figures

The four cartoons are reproduced by kind permission of the cartoonists.

Acknowledgements

This book grew out of an international conference on the Falklands Conflict, held under the aegis of the Department of International Relations at the University of Keele in September 1990. One of the aims of that conference was to bring together, not only different nationalities, but also different experiences: former participants – witnesses, as it were – civil servants and others still professionally involved, and analysts, academic and journalistic. The debate among these various species is reflected in the following pages.

For their advice and encouragement in the planning of the conference I am grateful to Professor Lawrence Freedman and to Peter Hennessy. Dr Anthony Seldon, Director of the Institute of Contemporary British History, was especially helpful at this early stage. Grants from the British Academy and the Foreign and Commonwealth Office secured the funding of the enterprise. Two of my undergraduate students, Robert Foot and Andrew Trehearne, acted as temporary unpaid administrative assistants and gave sterling service throughout, coping magnificently with every contingency and mixing easily with the great and the good around the conference table. Mrs Maureen Groppe provided invaluable secretarial support before and after the event. I should like to thank particularly a distinguished group of commentators on the original conference papers: Peter Hennessy, Michael Herman, Sir Rex Hunt, Admiral Sir Henry Leach, Dr Walter Little, Sir Anthony Parsons, Dr Oliver Ramsbotham, Professor Trevor Taylor, David Thomas and Robert Worcester. Professor Lawrence Freedman and Professor Alan James kindly agreed to take the chair at certain sessions, thereby giving everyone a respite from me.

There are perhaps two tests of a successful conference – that it is interesting and even enjoyable to attend; and that it has a long scholarly half-life. The Keele conference appears to have passed the first of those tests. It is about to take the second.

Alex Danchev
Keele and Oxford, 1991

Notes on the Contributors

Peter Beck is Reader in International History, Faculty of Human Sciences, Kingston Polytechnic. His numerous publications include *The International Politics of Antarctica* (1985) and *The Falkland Islands as an International Problem* (1988). He is a member of the Latin American Study Group of the Royal Institute of International Affairs, and has given expert evidence to the House of Commons Foreign Affairs and Education and Science Committees.

Christoph Bluth is Research Fellow in the Department of War Studies, King's College, London, where he specialises in arms control and Soviet strategy. His master's thesis at Trinity College, Dublin, was on 'Just War Theory and the Falklands/Malvinas Dispute'.

Peter Calvert is Professor of Comparative and International Politics at the University of Southampton. He is the author of *The Falklands Crisis: the Rights and Wrongs* (1982) and, with his wife Susan, who is a specialist on Argentine politics, *Argentina: Political Culture and Instability* (1989).

Alex Danchev is Professor and Head of the Department of International Relations at the University of Keele. A specialist in military history and international security issues, he has held fellowships in the Department of War Studies, King's College, London, the Woodrow Wilson Centre in Washington DC, and St Antony's College, Oxford. His latest work is a biographical study of Oliver Franks, *Founding Father* (forthcoming).

Clive Ellerby took a first class honours degree at the School of African and Asian Studies, University of Sussex, where he was awarded the Rose Prize as the best History finalist of 1986. In 1990 he completed his doctoral thesis at Exeter College, Oxford, on 'British interests in the Falkland Islands: economic development, the Falkland Lobby and the sovereignty dispute, 1945 to 1989'.

Lawrence Freedman is Professor and Head of the Department of War Studies, King's College, London, and one of Britain's premier strategic thinkers. Among many other works, he is the author of *Britain and the Falklands War* (1988) and, with Virginia Gamba-Stonehouse, *Signals of War: The Falklands Conflict of 1982* (1990).

Virginia Gamba-Stonehouse is Visiting Senior Research Fellow in the Department of War Studies, King's College, London. In 1983–4 she was Research Officer for the former Argentine Minister of Foreign Affairs, Nicanor Costa Mendez. She is the author of *The Falklands/Malvinas War: a Model for North-South Crisis Prevention* (1987), *Strategy in the Southern Oceans: a South American View* (1989) and, with Lawrence Freedman, *Signals of War: the Falklands Conflict of 1982* (1990).

Michael Herman, a graduate of the Joint Services' Staff College, was a British civil servant for thirty-five years. He was Gwilym Gibbon Research Fellow at Nuffield College, Oxford, in 1987–8, and is now an Associate Member of the College and an Honorary Senior Research Fellow of the Department of War Studies, King's College, London. He has published a number of articles on intelligence matters and is writing a book on the subject for the Royal Institute of International Affairs.

Guillermo Makin is the British correspondent of the Buenos Aires daily newspaper *Clarín* who specialises in Argentine politics and Anglo-Argentine relations. He organised several meetings of Argentine and British politicians in the aftermath of the 1982 war, and has given expert evidence to the House of Commons Foreign Affairs Committee.

Felipe Noguera is a partner in CEPPA (Centre for the Study of Applied Public Policy) in Buenos Aires, and in SOCMERC, the company responsible for extensive recent public opinion polling in Argentina. He has taken an active interest in the Falklands/Malvinas dispute, through work with the Fundacion Universitaria del Rio de la Plata.

Paul Rogers is Senior Lecturer in Peace Studies at the University of Bradford and a leading analyst of British defence policy.

He has given expert evidence to the House of Commons Foreign Affairs Committee, and to the unofficial Belgrano Enquiry, published in *The Unnecessary War* (1988) by the Belgrano Action Group.

Felipe Sanfuentes, a Chilean national, is an independent researcher specialising in the geographical and geostrategic problems of southern South America. A former international civil servant, his various appointments included service in Washington DC, and Central and South America. He now lives in Bethesda, Maryland.

Peter Willetts is Senior Lecturer in International Relations at the City University, London. He was Secretary of the South Atlantic Council from April 1985 to September 1988, and continues to be an active member and editor of the Council's series of Occasional Papers.

Introduction: A Matter of Life and Death

Alex Danchev

Scarcely is the blood dry before the ink begins to flow. The Argentine *coup de main* was delivered on 2 April 1982. The first British troops came ashore on 21 May, and took the Argentine surrender on 14 June. 'Still as Saxon slow at starting, still as weirdly wont to win.' *Puerto Argentino* reverted gratefully to its pre-war existence as plain old Port Stanley. *Las Islas Malvinas* gave place once more to the Falkland Islands. Within weeks, the first wave of Falklands literature, the great *tsunami* of 1982–3, came crashing down.[1] It was a potent mixture of eye-witness report and deep-throat history, unavoidably reliant on inside information conveyed orally, hurriedly, and as a rule unattributably. The characteristic first-wave work is compelling, partial, knowledgeable, and raw. It reeks of authenticity but will not pause for breath. Like all historical writing, it tells of its times. Nor is this a negligible achievement. Indeed, the best of these works are in many ways unsurpassed. *The Battle for the Falklands,* by the British journalists Max Hastings and Simon Jenkins, remains one of the most incisive and comprehensive accounts of the conflict, quarried annually for its lessons by a fresh batch of cadets at the US Naval Academy.[2] Others offer important keys to the wonderland world of the military junta in Buenos Aires. *Falklands – the Secret Plot* (originally *Malvinas – la trama secreta*), by the Argentine journalist Oscar Cardoso and his colleagues, contains a transcript of the delayed, fractured, comic-opera telephone conversation between two Presidents, Galtieri and Reagan, variously out of touch on the very eve of calamity in April 1982. The leitmotif of their exchanges – the insistent question of one to the other, 'Do you understand me, Mr President?' – is a mournfully apt commentary on the matter of life and death on 'that little ice-cold bunch of land down there', as Ronald Reagan once described it.[3]

Ten years on, the second wave has broken. It seems that the tributary of personal reminiscence from high and low (but chiefly high) has not run dry after all. Politicians grow more garrulous as soon as they leave office – or are forced out. From the British top table we have the Carrington memoirs, the Whitelaw memoirs, and even that most unlikely artefact, the Tebbit memoirs.[4] Tremulous, we await the Thatcher version. From Washington come the Haig memoirs and the Weinberger memoirs, and a shoal of smaller fry.[5] There is the fascinating collective testimony elicited by Michael Charlton for a series of BBC radio programmes, now a book, called *The Little Platoon*, a phrase appropriately culled from Edmund Burke:

> To be attached to the sub-division, to love the little platoon we belong to in society, is the first principle (the germ as it were) of public affections, the first link by which we proceed towards a love to our country and to mankind. The interest of that portion of social arrangement is a trust in the hands of all those who compose it; and as none but bad men would justify it in abuse, none but traitors would barter it away for their own personal advantage.[6]

Of the diplomats, Our Man in Buenos Aires, Sir Anthony Williams – the author of a prophetic complaint about the typical British approach to the Falklands, 'to have no strategy at all beyond a general Micawberism' – recorded some candid interviews before his early death in 1990.[7] Our Men in Washington and at the UN, Sir Nicholas Henderson and Sir Anthony Parsons, both writers of distinction, have published revealing accounts of their stewardships.[8] Our Man in Lima, Charles Wallace, bids to do the same.[9] These British accounts can be matched interestingly against the waspish reflections of Jeane Kirkpatrick, the dissentient US Representative at the UN, who could not conceive that Margaret Thatcher, 'a decolonizer in a long line of decolonizers', would or should go to war over the misbegotten Falkland Islands. When war came, which way would Washington jump? American action, or inaction, was an issue of great moment for both protagonists, as Christoph Bluth makes plain his treatment of Anglo-American relations in the conflict, in Chapter 10 of this book. For her part, Kirkpatrick had a distressingly clear view of the mainspring of US foreign policy:

> I thought the US should attempt mediation. If the mediation failed, then US interest, I thought, dictated that we should remain neutral: because the US had a continuing interest in good relations with Latin America as well as with the UK; because Britain herself made clear that she saw the NATO commitment as limited to the NATO area, and because Britain did not stand with the US elsewhere; and finally, because under existing arrangements, the British already enjoyed the tangible benefits that the US would in any case make available: we shared intelligence and weapons with the UK.[10]

Unhappily for Argentina, this was not the view that prevailed in Washington.

It was only to be expected that the perspective from Buenos Aires is rather less well represented, especially in English-language publications. The victor often lays first claim to the plaudits of history. Foreign Minister Costa Mendez has entered the literary lists, but as yet only in piecemeal fashion. His memoirs are apparently now complete. One can anticipate a closely argued but largely unreconstructed *apologia pro vita sua.* On the decision to invade, for example, he has recently argued that 'the British Government took advantage of the Davidoff presence in the [South] Georgia Islands to exaggerate the incident and mount a scheme that should lead to the rupture of negotiations and the building up of Fortress Falkland. Argentina had no other choice but to re-take the islands to oppose the British decision.'[11] Self-exculpation is an understandable refuge from the ignominy of the outcome, but some readers will wince at the use of the term 're-take', with its smuggled implication that the original sin was British. History is summoned to account in a sovereignty dispute. The past is pliable. In Chapter 1, Peter Beck demonstrates with great forensic skill the truth of Sir Anthony Parsons's magnificent quotation from T.S. Eliot: 'History may be servitude, History may be freedom'.[12] In this instance, History tends to contradict the Argentine Foreign Minister. If there was one thing that the benighted British Government did *not* want in March 1982, it was the rupture of negotiations and the building up of Fortress Falklands (an expression then unknown). As Costa Mendez himself has written, with both economy and justice, at that juncture 'London had only one plan: procrastination'.[13]

There is not much in the present volume about the war itself. Paul Rogers analyses the consequences for British defence policy in Chapter 9, but is more concerned with dispositions than operations. The course of the campaign is well-known, and there is now a wealth of material on the tactical and technical aspects, a surprising amount of it generated by the combatants themselves. The generals and admirals of both sides have been remarkably forthcoming – in one case from prison, in the other from retirement: such are the spoils of war.[14] The poor bloody infantry, Argentine and British, have been induced to open their hearts on film and in print, most notably in the rich narrative tapestries of Peter Kosminsky and Martin Middlebrook.[15] The testimony they offer is wonderfully sobering. There is no thirst for blood and little talk of honour. The Argentine army officer who told Middlebrook in 1987 that he hoped 'one day [to] fight again in the Malvinas with better equipment, better training, and settle the debt for those of my men who died there' doubtless still represents a certain emotional tendency, but it appears to be an increasingly inadmissible one.[16] In 1991, in the Gulf War, elements of the two navies – many of them wearing the medals of the earlier war – found themselves on the same side, operating together in the multinational force ranged against Saddam Hussein. 'The Argies are doing a grand job', reported a Royal Navy spokesman sportingly. Argentine officers were even heard to refer tactfully to 'the Falklands', rather than 'the Malvinas', in conversation with their British opposite numbers.[17]

Since the mid-1980s, in fact, the dominant spirit among the soldiery – victor and vanquished alike – has been unmistakable. It is one of profound regret. Too many of their fellows came to the islands and died: 655 Argentine and 255 British servicemen. Cumulatively and inchoately, the testimony of these seasoned youngsters has managed to convey something of the unutterable horror of war. In the South Atlantic the usual disjunction between expectation and reality was viscerally felt, especially among the unblooded, unprepared Argentine conscripts. (It is also painfully evident in the posthumously published letters of Lieutenant David Tinker RN, written from his berth on *HMS Glamorgan* as the Task Force wended its inevitable way south.)[18] Neal Ascherson has called this agonising feeling a disease, *solitudo superstitis*, the loneliness of the survivor – 'the failure to make "sense" of an experience, the bitter realisation that what has happened to you is incommunicable except to another survivor'. Paul Fussell has taught us to recognise it as irony.[19] The Falklands War offers a poignant demonstration of the witty but dispiriting proposition that all wars are ironic because all wars are worse than anticipated. Thus an Argentine private reflected: 'When we first arrived in the Malvinas, the general feeling was that the war – such as it was – was over; it had been won for Argentina on 2 April'.[20] As Clausewitz remarked long ago, the aggressor is always peace-loving.

There is not much here, either, about the infamous events at sea which brought home to so many people the sudden nemesis of war: the sinking of the *General Belgrano* on 2 May 1982 and *HMS Sheffield* two days later, the last moments of both ships permanently frozen in haunting photographs. Such events – spectacular, shocking, solemn – crystallise the image and help to define the character of a conflict. The emotional pitch they create cannot be sustained, but is recreated and redefined in every war – at a somewhat similar stage in the Gulf War, by the precision bombing of hardened targets (command bunkers, missile silos and the like) in Iraq. For almost a month the world goggled dumbly at such precocious technical virtuosity, as video film of the bombs' delivery was replayed in slow-motion on the nightly news. Then, on 13 February 1991, nemesis struck. A bunker in a Baghdad suburb had been used as a population shelter: hundreds of Iraqi civilians were killed. Reality obtruded like a severed limb. 'Already,' wrote J.G. Ballard, 'one can visualise the combatants in a future war returning from their sorties and firefights to scan the evening rushes, and perhaps planning the next day's tactical strike in terms of its viewer-potential.'[21] As the armaments manufacturers have discovered, 'surgical strike' makes a terrific video game. War is the ultimate snuff movie. More than ever, it is necessary to press the pause button and reflect. This book is an attempt to do just that.

Unusually, 'the Belgrano affair' lingered on long after the war was over. As late as 1988 a number of concerned citizens calling themselves the Belgrano Action Group were attempting to reveal the hidden motive and thus pin the political blame for the sinking. They concluded equivocally:

It remains unclear whether the Government was merely guilty

> of incompetence in failing to protect the islands before the Argentine attack and so involving Britain in an avoidable war, or whether they also deliberately rejected the possibility of a peaceful outcome, preferring a military solution.[22]

Cock-up or conspiracy? The issue is examined by Lawrence Freedman in Chapter 8 in the context of the over-familiar but under-analysed concept of escalation.[23] Freedman introduces the helpful notion of the just escalation (by analogy with the well-established doctrine of the just war). In these terms the sinking of the *Belgrano* has been frequently represented as an unjust escalation. Yet the more we find out about the British decision, the more it seems that the military arguments were sufficient. In the summary judgement of Admiral Sir Henry Leach, then Chief of the Naval Staff, '*if* the sinking of the *Belgrano* is to be regarded as a classic escalation . . ., responsibility for that "escalation" properly lies with the Argentines for positioning a powerful force so that it constituted a threat which could not prudently be ignored'.[24] This brings us to the difficult art of threat assessment.[25] Some fundamental issues in 'long-range' threat assessment are discussed with sensitivity and authority by Michael Herman in Chapter 7, starting with the central question of warning (or lack of warning) of any surprise attack – the kind that took place on 2 April 1982, for example. In this perspective Herman rightly calls attention to the unprecedented openness of the Franks Report,[26] which can indeed be read as a treatise on long-range threat assessment and the difficulties of interpreting Argentine intentions. In the intelligence field as in others, that remarkable document must be counted the supreme revelation of the first wave of Falklands literature and a prime source for the second. Its message and purpose are re-examined by the present author in Chapter 6, drawing for the first time on the evidence of a number of those principally involved, including the legendary Lord Franks himself.

More immediate threat assessment tends naturally to collapse intentions into capabilities. The mere presence of an Argentine cruiser could not prudently be ignored, regardless of whether it was closing on the Task Force or steaming away from it – the occasion of much retrospective recrimination. Like the grand fleets of old, 'in being', it constituted a threat. For the British, time was pressing – the shadowing submarine might lose it –

and anyway no reliable inference could be drawn from its direction of travel. 'The target's course was completely irrelevant,' Leach has written dismissively; 'no professional commanding officer operating in a threat-area will maintain one course for longer than, at most, three minutes and will then make an alteration of some sixty degrees.'[27] The fog of war – or rather the fog of diplomacy – made up the account. We are inclined to forget that on the day of decision (2 May 1982), before the fighting had properly begun, no one in London could be confident that a favourable 'military solution' was remotely possible of achievement. It is certainly true that the Belgrano affair is of great political interest – much greater than the Thatcher Government's apologists would have us believe – but the principal part of that interest is domestic and *post facto*, a consequence of persistent official economies with the truth and a quixotic gesture by the Ministry of Defence whistle-blower Clive Ponting.[28] During the war there was neither a cock-up nor a conspiracy, nor even a conspiratorial cock-up (often the most persuasive variant). The most discreditable aspect of the whole affair was the subsequent attempt to conceal the facts of the engagement – concealment far in excess of the exiguous requirements of national security. That is in itself a serious matter. By comparison with the other charges, however, it is no more than an egregious case of unjust dissimulation.

One glaring deficiency of the first-wave literature is unfortunately perpetuated in much of the second: a certain lack of regard for the interactive nature of the conflict.[29] Actions have consequences in the international arena, especially in cases of dispute. Moves prompt counter-moves; anticipation urges prevention. In this particular case, it is clear that British actions were monitored much more closely by Argentina than were Argentine actions by Britain, an imbalance of attention reflected in Lord Carrington's famous remark that the Falklands ranked number 249 on his list of priorities as incoming Foreign Secretary in 1979. In the collective psyche of policy-makers in Argentina, on the other hand, the repossession of the Malvinas had both geostrategic and totemic significance: it was inescapably a high priority, and increasingly an urgent one, as Peter Calvert and Virginia Gamba-Stonehouse explain in Chapters 2 and 5 respectively. Thus imbalance of attention testified to an imbalance of interest. For Buenos Aires, the stakes were axiomatically high. For London, as long as the Argentines (and for that matter the islanders) were quiescent, the stakes were exceedingly low. They were raised at a stroke by the Argentine *coup de main* – the fundamental (and paradoxical) interaction of the conflict. Such a gross imbalance could only exacerbate the endemic problem of mutual incomprehension. In diplomacy as in life, it takes two to tango. As Virginia Gamba-Stonehouse makes clear, it seemed that the British, unbalanced by the tempo, never mastered the steps.

There were other interactions, and further complications. This was not a purely Anglo-Argentine conflict. It had a Chilean dimension, usually unacknowledged, often forsworn, but discussed here with some passion by Felipe Sanfuentes, himself a Chilean, in Chapter 3. More important still, had the Foreign Office indicated enthusiasm for anything faster than a stately minuet, it was assumed in Buenos Aires and even London that the islanders or their more thuggish supporters were always ready to cut in. In these quarters diplomatic sweet-talk was considered quite disgraceful. It was but a short step to the perception that British foreign policy was hostage to a small group of die-hard kelpers. In this context the notorious Falkland Lobby has been endowed with almost supernatural powers of persuasion – or at least veto. Scapegoat or saviour, according to taste, 'the Lobby' is indeed a peculiar phenomenon, though by no means the almighty presence suggested by the demonology. It is expertly anatomised in Chapter 4 by Clive Ellerby, who utilises a variety of hitherto untapped documentary sources to provide the most complete study of that gryphon-like pressure group yet to appear.

The remaining contributions look to the present and to the future. In Chapter 11, Guillermo Makin charts the tough learning process that was the last decade of Anglo-Argentine diplomacy. He speculates on the opening of new windows of opportunity, and wonders wryly whether either country will have the heart to jump through them. In the final chapter Felipe Noguera and Peter Willetts are perhaps a shade more positive, on the basis of a careful analysis of their fascinating recent exercise in public opinion polling in both countries. The old warriors have been pushed aside – Galtieri in 1983, Thatcher in 1990. It appears that there may well have been a hopeful shift in public attitudes. The Cold War has thawed; in

Europe, everything is possible once more. Is this not a new era, as some Argentine voices have proclaimed?[30] At once a note of caution sounds. Stuck fast in the new era lies the old dispute, unregenerate and unresolved, a mammoth that has failed to evolve: the question of sovereignty. Sir Anthony Parsons, for one, has argued that 'there is no definitive solution available for the time being'.

> Neither side will modify its present position and no British Government, present or future, will again leave the islands inadequately garrisoned against surprise attack. The best that can be hoped for is (a) that democracy will take ineradicable root in Argentina; this will over the years allay some of the apprehensions of the islanders; (b) that functional co-operation between the islands and the mainland will be resumed in terms of sea and air communications, joint economic ventures, educational and health facilities, etc; and (c) that Anglo-Argentine relations will be conducted as far as possible separately from the dispute: the resumption of diplomatic relations [in 1990] is a good omen in this context. Time, long time, must be allowed to do its work on the attitudes of all concerned.[31]

This is surely realistic. In a profound sense it is also optimistic – as Edmund Burke reminds us, 'history consists, for the greater part, of the miseries brought upon the world by pride, ambition, avarice, revenge, lust, sedition, hypocrisy, ungoverned zeal, and all the train of disorderly appetites, which shake the public with the same

> troublous storms that toss
> The private state, and render life unsweet'.[32]

NOTES

1. Literary 'waves', and a number of other ideas in this introduction, are adapted from my essay 'Life and Death in the South Atlantic', *Review of International Studies* 17 (1991). For contemporary surveys of the first

wave see Neal Ascherson, 'By San Carlos Water', *London Review of Books*, 18 November–1 December 1982; Simon Collier, 'The First Falklands War?' and Lawrence Freedman, 'Bridgehead Revisited', *International Affairs* 59 (1983) pp. 459–64 and 445–52; Geoffrey Wheatcroft, 'The Fighting and the Writing', *Times Literary Supplement*, 13 May 1983. For later assessments see Eduardo Crawley, 'The Paradox Islands', *Times Literary Supplement*, 25–31 March 1988; and Lawrence Freedman, 'The Literature on the Conflict', in *Britain and the Falklands War* (Oxford: Blackwell, 1988) pp. 122–7, or *Contemporary Record* 1 (1987) pp. 34–5.

2. Max Hastings and Simon Jenkins, *The Battle for the Falklands* (London: Pan, 1983).
3. Oscar R. Cardoso et al., *Falklands – the Secret Plot* [1983] (East Molesey: Preston Editions, 1987) pp. 83–6. The book's title is somewhat misleading: 'plot' here means storyline.
4. Lord Carrington, *Reflect on Things Past* (London: Fontana, 1989); Lord Whitelaw, *The Whitelaw Memoirs* (London: Aurum, 1989); Norman Tebbit, *Upwardly Mobile* (London: Futura, 1989).
5. Alexander Haig, *Caveat* (London: Weidenfeld and Nicolson, 1984); Caspar Weinberger, *Fighting for Peace* (London: Michael Joseph, 1990); John Lehman, *Command of the Seas* (London: Macmillan, 1988); Bob Woodward, *Veil* (London: Headline, 1987) [for William Casey]; Michael Charlton, *The Little Platoon* (Oxford: Blackwell, 1989) [for Thomas Enders and Vernon Walters].
6. Edmund Burke, *Reflections on the Revolution in France* [1790] (London: Penguin, 1986) p. 315. See Charlton, p. 1.
7. Sir Anthony Williams in Charlton, pp. 125–32, and in Michael Bilton and Peter Kosminsky, *Speaking Out* (London: Grafton, 1990) pp. 29–35. For the complaint, see Williams to Fearn, 2 October 1981, quoted in Cmnd. 8787, *Falkland Islands Review* [the Franks Report] (London: HMSO, 1983) p. 30.
8. Sir Nicholas Henderson, 'America and the Falklands', *The Economist*, 12 November 1983; Sir Anthony Parsons, 'The Falklands Crisis in the United Nations', *International Affairs* 59 (1983) pp. 169–78. See also their subsequent reflections, 'The Washington Embassy', *Diplomacy and Statecraft* 1 (1990) pp. 40–8, and 'The Future of the Falklands', *International Relations* IX (1987) pp. 131–9.
9. Charles Wallace, unpublished memoirs.
10. Jeane J. Kirkpatrick, 'My Falklands War and Theirs', *The National Interest* 18 (1989/90) pp. 11–20 (quotations pp. 11 and 17); excerpted in 'The Falklands – a war the Americans lost', *Sunday Times*, 17 December 1989.
11. Nicanor Costa Mendez, 'The Malvinas Conflict in the South Atlantic Context', paper delivered to the Pugwash Conference on Security and Co-operation in the South Atlantic, Buenos Aires, 28–31 October 1990.
12. Sir Anthony Parsons, comment on Peter Beck's paper (Chapter 1 below) for the Conference on the Falklands Conflict, University of Keele, 27–8 September 1990.
13. Costa Mendez, op. cit.
14. See Admiral Lord Lewin, evidence to House of Commons Foreign Affairs Committee, Third Report, Session 1984–5, *Events Surrounding the Weekend of 1-2 May 1982* (London: HMSO, 1985); Major General Sir

Jeremy Moore and Rear Admiral Sir John Woodward, 'The Falklands Experience', *Journal of the Royal United Services Institute* 128 (1983); Major General Julian Thompson, *No Picnic* (London: Leo Cooper, 1985). Lawrence Freedman and Virginia Gamba-Stonehouse make extensive use of high-level Argentine military testimony in *Signals of War* (London: Faber, 1990), as does Martin Middlebrook (see note 15).

15. Peter Kosminsky, 'Falklands – the Untold Story', Yorkshire TV documentary (1987); Martin Middlebrook, *Task Force* and *The Fight for the 'Malvinas'* (London: Penguin, 1987 and 1990). I have analysed Middlebrook's work more fully in the *Journal of the Society for Army Historical Research* LXVIII (1990) pp. 267–9.
16. Quoted in Middlebrook, *Malvinas*, p. 291.
17. Kathy Evans, 'Gulf between old enemies is ended', *Guardian*, 14 February 1991.
18. Hugh Tinker, ed., *A Message from the Falklands* (London: Junction, 1982).
19. Neal Ascherson, 'Cain finds pity in the Falklands', *Observer*, 5 April 1987; Paul Fussell, *The Great War and Modern Memory* (Oxford: Oxford University Press, 1975).
20. Quoted in Middlebrook, *Malvinas*, p. 290.
21. J.G. Ballard, 'Old bloodshed, as if in a dream', *Guardian*, 28 February 1991.
22. Belgrano Action Group, *The Unnecessary War* (Nottingham: Spokesman, 1988) p. 178.
23. See also his 'Escalators and Quagmires', *International Affairs* 67 (1991) pp. 15–31.
24. Admiral Sir Henry Leach, comment on Lawrence Freedman's paper (Chapter 8 below) for the Conference on the Falklands Conflict, University of Keele, 27–8 September 1990. The best case for the prosecution is developed in Arthur Gavshon and Desmond Rice, *The Sinking of the Belgrano* (London: Secker and Warburg, 1984). Simon Jenkins, 'The Truth about the Belgrano', *The Spectator*, 11 June 1983, is a peerless first-wave commentary.
25. A subject on which the participants at the Keele conference had the benefit of the accumulated wisdom of several former practitioners. See Peter Hennessy's report of the proceedings, 'Whitehall Watch', *Independent*, 8 October 1990.
26. Originally released in 1983, now to be republished, with an introduction by Alex Danchev (London: Pimlico, 1992).
27. Leach, op. cit.
28. See Peter Greig [a pseudonym], 'Revelations', *Granta* 15 (1985) pp. 252–61; Richard Norton-Taylor, *The Ponting Affair* (London: Cecil Woolf, 1985); Clive Ponting, *The Right to Know* (London: Sphere, 1985).
29. The outstanding exception is Freedman and Gamba-Stonehouse, *Signals of War*, which is centrally concerned with exactly this issue.
30. See for example Ambassador Mario Campora, 'A Special Relationship between Argentina and the United Kingdom: is it possible?', speech made at the Club Argentino, London, 14 November 1990, published and distributed by the Argentine Embassy in London.
31. Parsons comment, op. cit.
32. Burke, pp. 247–8, quoting Spenser's *Faerie Queene*.

1 The Policy Relevance of the Falklands/Malvinas Past

Peter Beck

Speaking for the British government in April 1982 at the time of an escalating Argentine threat to British control over the Falkland Islands, Sir Anthony Parsons informed the UN Security Council that 'my Argentine colleague and I could debate endlessly the rights and wrongs of history, and I doubt whether we would agree'.[1] Nicanor Costa Mendez, the Argentine Foreign Minister, proved the point when asserting that there existed no cause for disagreement:

> The representative of the United Kingdom said that he had doubts about being able to arrive at an agreement with the representative of my country as to the historical vicissitudes. This is possible, but it would seem difficult for us not to agree on the facts of history which are absolutely indisputable.[2]

Argentina proceeded to occupy the disputed islands and gave effect to its 'indisputable' historical and legal rights to sovereignty over a territory depicted hitherto on Argentine maps in an act of cartographical wish-fulfilment as the *Islas Malvinas.*

At the time the Argentine people believed that – to quote Roberto Guyer, an Argentine diplomat involved in various stages of Anglo-Argentine exchanges since the late 1960s – 'going back in history, the Malvinas were Argentina's, are Argentina's. So, for us it is a question of national honour and national dignity'.[3] Subsequently one Argentine conscript looked back to the 1982 war:

> We'd been lectured a lot about the Malvinas, the importance of their recovery . . . they talked a lot about the English as invaders of something that is ours. We felt that we were going to the Malvinas to defend something that was ours.[4]

Inevitably, the invasion of April 1982 undertaken to terminate Britain's 'illegal' occupation allowed the islands to be renamed on-the-spot as the *Malvinas*: thus, the military action was interpreted as giving substance to Argentina's historical rights to the islands, which were depicted in a popular wartime song as '*Las Hermanitas Perdidas*' ['the lost little sisters'] of the Argentine family. During the war a book by Admiral Destefani was translated and widely distributed throughout the world 'to clarify the Argentine truth and make it known everywhere':

> Thus it may be possible for people . . . to better understand how substantial our rights are. The simple geographical, historical and legal truths constitute the best defence of our rights of sovereignty over the three southern archipelagos (the Malvinas, South Georgia and South Sandwich Islands) . . . The fighting still continues . . . Whatever the cost may be and however long it may take, the three archipelagos must be ours, because our cause is just. The Malvinas are Argentine. For historical reasons . . . for geographical reasons . . . for reasons of international law . . . And because, from 1833 onwards, which was the year in which we were attacked, we have never given them up, nor will we ever do so![5]

By contrast, in London the Argentine action was viewed as an illegal invasion of British territory infringing historical and legal rights reinforced by a continuous occupation spanning a period of nearly 150 years. On 3 April 1982 Margaret Thatcher summarised her government's position:

> British sovereign territory has been invaded by a foreign power . . . the lawful British government of the islands had been usurped . . . I must tell the House that the Falkland Islands and their dependencies remain British territory. No aggression and no invasion can alter that simple fact . . . Argentina has, of course, long disputed British sovereignty over the islands. We have absolutely no doubt about our sovereignty, which has been continuous since 1833.[6]

The despatch of the Task Force, though prompted by a range of factors, was presented to domestic and international audiences primarily in terms of the need to protect legitimate British rights through the recapture of the islands.

In April 1990 Norman Stone, Regius Professor of Modern History at Oxford, justified his new role as a *Sunday Times* columnist on current affairs in terms of the fact that 'this is precisely what a professor of modern history ought to do: to set contemporary matters in a historical perspective'.[7] History provides an invaluable framework of reference for those seeking to understand the present and to discuss future policy possibilities. Certainly, it is difficult to look forward in an informed manner without knowing exactly where we are at the present time, how we got there, and the real nature of the existing situation. We need history to make sense of contemporary debates as well as of future possibilities. This recalls the farewell lecture of Stone's predecessor, Michael Howard, on history's value to society. 'From that study we learn what we have been, understand what we are, and gain intimations of what we might become.'[8]

The craft of history stresses the need for objectivity and the study of the past for its own sake, but in reality history is often exploited for functional reasons related to the desire to understand the background to the contemporary world. In certain instances a more subjective approach has been adopted according to which the selection and interpretation of evidence on a topic is often influenced by the desired result, and particularly by an effort to impart an added credibility through the establishment of a meaningful continuity with the past. At times considerable distortion and manipulation of the 'facts', in conjunction with the perpetuation of historical amnesia on inconvenient points, has occurred in order to meet the desired objective; indeed, Eric Hobsbawm has employed the phrase 'the invention of tradition' to describe the frequent creation of a past which never existed.[9]

There is no space to pursue this point in detail, but numerous examples of the functional past can be identified, most notably, the use of history to justify and support either nationalism or the Soviet regime. One of the major compliments paid to historians was Nikita Khrushchev's statement that 'historians are dangerous and capable of turning everything topsy-turvy'. Perhaps he was right, but this observation establishes the perceived significance of history in a state where the subject has been employed, indeed exploited, for policy reasons, to promote the legitimacy of the regime, to stress the importance of class, and to foster the integration of non-Russian nationalities into the Soviet bloc.[10] However, recent events in Central and Eastern Europe suggest that a class-based history proved unsuccessful in overriding the national identities of the Estonians, Lithuanians, Poles and others as compared to the more enduring force of the histories and mythologies utilised to develop and sustain a national consciousness. For example, Lech Walesa's recent memoir, recording his initial dismissal of the abstract and unreal nature of a denationalised, class-based history taught during his schooldays, articulated his eventual appreciation of the functional value of history in terms of serving Polish national interests.[11] Within the USSR an active debate about history proved one of the major consequences of Mikhail Gorbachev's policy of *glasnost* as Soviet citizens sought to reinterpret their past, to rethink the principles and practice of the current regime, and to appreciate that 'the debate about the past is also a debate about the future of Soviet society'.[12] Recent controversies about the proposed school curriculum in history, most notably concerning the role of national history, have illuminated the perceived political significance of the subject in Britain.

Governments have often exploited history for functional reasons, and one history professor has complained that 'there is no limit to the historical nonsense that is believed especially when propagated by professors of history when they venture to make history serve the interests of the state'.[13] The subject, though not synonymous with propaganda, frequently gives this impression, since each country writes, and then rewrites, its own history according to individual prejudices and requirements.[14] In such circumstances, the historian's search for objectivity, including his role in disrupting and exposing the mythologies of functional histories, may be compromised by the demands made upon his subject, especially as the imposition of a sceptical mind may prove both difficult and unwelcome within official circles. Therefore, history, though capable of exposing false versions of the past, has often been used, or rather abused, to construct and strengthen such false accounts. During the Falklands War of 1982 the dual role of the discipline was articulated by David Tinker, a young naval lieutenant stationed in the South Atlantic on *HMS Glamorgan*, in a letter received in England nine days after his death when the ship was hit by a Exocet missile:

> We had just been attacked by their Mirages [and] we were saying to each other that it couldn't be real. Why were we fighting Argentina? And who wanted to live here in the Falklands anyway? I cannot think of a single war in Britain's history which has been so pointless . . . What Mrs Thatcher does not realise is that the Argentinians *believe* that the Malvinas are theirs . . . The bravery of all their pilots shows that they are more than 'mildly' interested in the Falklands . . . The professional forces of both sides do what they are told. So if two megalomaniac idiots tell them to beat each other's brains out, they do . . . But enough of my sarcasm . . . They really should not send people in the Services to study history at university. The Captain just says things about democracy, and duty, and resisting invasion.[15]

Michael Akehurst, an international lawyer, claimed that 'it is doubtful whether any single word has ever caused so much intellectual confusion and international lawlessness' as that of sovereignty, and his studies of the Anglo-Argentine dispute regarding the question of which government possesses the exclusive right to exercise the functions of the state in the Falkland Islands merely reinforced this impression.[16] Both disputants, attaching considerable significance to the presentation and relative strength of their respective historical and legal cases, view the sovereignty question in rather dogmatic, black-and-white terms [Figure 1].[17] For example, in May 1982 Francis Pym, the British Foreign Secretary, asserted that 'we are not in any doubt about our title to the Falkland Islands' in spite of the fact that 'Argentina remains so obdurate in upholding a claim which it believes is valid but which we are confident is invalid'.[18]

In general, the Falkland islanders, refusing to consider the validity of an alternative Argentine version of the past, have assumed an even more dogmatic view of the question, as stated by Sydney Miller, a retired sheep station manager and editor of *The Falkland Islands Journal*:

> My family has been here since 1860 and my sons are fourth generation Falkland Islanders. I probably know more of the true facts of the history of these Islands than anyone else as I have made it my business to ascertain the true facts of the

MALVINAS ISLANDS

Let the American People be the Judge

ARGENTINA'S POSITION

Geography

The Malvinas Islands are located on the South Atlantic continental shelf, 350 miles off the Argentine Coast.

History

The Malvinas were declared Spanish possessions by the Papal Bulls "Inter Coetera" and "Dudum Si Quidem" in 1493; by the Treaty of Tordecillas in 1494 and the Papal Bull "Ea Quae" in 1506.

Great Britain acknowledged Spanish sovereignty over the Islands under the Treaty of Madrid (1667-1670), the Treaty of Utrecht (1713) the Treaty of Madrid (1715-1758), the Treaty of Aquisgran (1748); the Treaty of Paris (1763); The Peace of Versailles (1783) and the Treaty of San Lorenzo (1790).

For 57 years, starting in 1774, Spain and then Argentina governed the Malvinas during which time Great Britain made no claims on the Islands. In 1825 when Argentina and Great Britain signed a treaty of friendship, mutual commerce and navigation, there was still no claim made on the Malvinas by Great Britain.

There have been 19 Spanish governors in the history of the Malvinas and 5 Argentine until the British Invasion of 1833.

In that year, the British expelled the Argentine governor and Argentine inhabitants without asking whether they wished to remain, either as Argentine citizens or British subjects.

Today after having re-populated the island with people of British origin, Britain manifests concern over the rights of these more recent inhabitants to remain as Argentines or Britons.

From 1833 on, Argentina has tried unsuccessfully with Great Britain and before all international organizations, to reclaim the Malvinas. For the past 17 years, the United Nations has urged negotiations between both countries to settle ownership claims. Great Britain however, has evaded and eluded such negotiations.

Meantime, for the inhabitants of the Malvinas, Argentina has contributed to a progressively higher standard of living by improving health and sanitation services, sea and air transportation with the mainland, food and fuel supplies and educational opportunities.

These efforts by Argentina for the inhabitants of the Malvinas resulted in a commendation by the United Nations. (Resolution 3160, dated December 14, 1973).

In February 1982, Argentina brought maximum pressure to bear on an attempt at settlement with no response from the British government. Instead British warships were sent to the South Georgia Islands, dependencies of the Malvinas, to expel Argentines, working there under an agreement by both countries.

GREAT BRITAIN'S POSITION

Geography

History

Invasion and taking of the Malvinas by force on January 3, 1833 expelling both the Argentine authorities and inhabitants.

APRIL 2, 1982

After 149 years of Argentina's claim of sovereignty and 17 years of negotiations which Great Britain rejected or evaded, Argentina has recovered the Malvinas without casualties to a single British inhabitant or soldier.

TODAY

The world stands perplexed at a spectacle seemingly out of the 18th Century: A naval fleet belonging to an colonialistic country sent to attack an American country in an effort to re-establish colonization.

CITIZENS OF ARGENTINA – 35 Sutton Place, New York, N.Y. 10022

FIGURE 1.1 *A black-and-white view of the Malvinas past (New York Times,* 18 April 1982)

> early discovery by British seamen and certainly the facts of the absurd and unfounded Argentine claims . . . We totally British people throughout the Islands have had to listen to so much in the way of broadcasts by apparently senior people in London who to too great an extent get their facts very wrong.[19]

Inevitably, opinion polls establish the islanders' deep commitment to a future British connection.[20]

'The facts' vary. The constant references to historical 'truths' characteristic of Argentine accounts, like that of Destefani, typify the strong position taken within Argentina, where there exists a tendency to dismiss the British position as resting on nothing more than the exercise of force and the continuation of an illegal colonial presence. In this connection, it is worth noting the 'juridical culture' allegedly influential in Argentina – there exists a preoccupation with legal rights and controversies regarding territorial rights – as well as understanding 'how much importance the Latin American legal mind and, often, governmental attitudes, attach to historical antecedents.[21]

Significantly, the British House of Commons Foreign Affairs Committee devoted considerable time to historical aspects during its post-1982 war enquiry into the Falklands question.

> An assessment of the claims of Argentina and the United Kingdom depends critically on their interpretation of the early history of the Islands, particularly between 1811 and 1833, and of the legal significance of the acts of commission and omission of the two Governments during this period and subsequently.[22]

Similarly, Jeane Kirkpatrick, the USA's UN representative during the 1982 war, identified the utility of history in comprehending this 'extraordinarily complex' controversy:

> We have come to appreciate how deep the roots of the conflict are. Britain [has been] in peaceful possession of the Falkland Islands for 150 years . . . Yet we know too how deep is the Argentine commitment to recover islands they believe were taken from them by illegal force. This is not some sudden passion, but a long-sustained national concern that also stretches back 150 years.[23]

Clearly both governments agree about the importance of the historical dimension, but disagree regarding the precise nature and interpretation of the Falklands/Malvinas past. The existence of rival versions of the history of the islands (Figure 1.2) remains a principal source of Anglo-Argentine controversy, especially as both disputants display confidence in the rightness of their respective versions of 'history', which have proved a function of the demands of politics and law rather than of those of historical scholarship. History has often been used and abused; indeed, the constant references to historical 'truths' reveal the unhistorical nature of this process, since no responsible historian accepts the possibility of ever arriving at a definitive version of the past. The resulting impasse, both causing and surviving the 1982 war, merely exacerbates the prospects for a resolution of the dispute.

Against this background, the study of the Falklands/Malvinas past provides a strong foundation for an informed evaluation of the causes and nature of the Anglo-Argentine dispute, while offering also an invaluable contribution to any discussion regarding the likely course of future developments and the feasibility of various policy options. This chapter will focus primarily upon the historical background to the sovereignty dispute, but it needs to be remembered that wider considerations help to explain the manner in which the dispute developed and escalated into war during 1982. For example, one can point to the traditional Argentine exaggeration of the country's role on the international stage, to Argentina's misreading of the pre-1982 lessons of history, and to Britain's longstanding reluctance to acceptance its declining world role. Speaking at Cheltenham a few weeks after Britain's 'great victory' over Argentina, Margaret Thatcher drew freely upon history to present her view that the so-called 'spirit of the South Atlantic' embodied the 'real spirit of Britain'.

> When we started out there were the waverers and the fainthearts . . . Those who believed that our decline was irreversible – that we could never again be what we were . . . Well they were wrong. The lesson of the Falklands is that Britain has not changed and that this nation still has those sterling qualities which shine throughout history. This generation can match their father and grandfathers in ability,

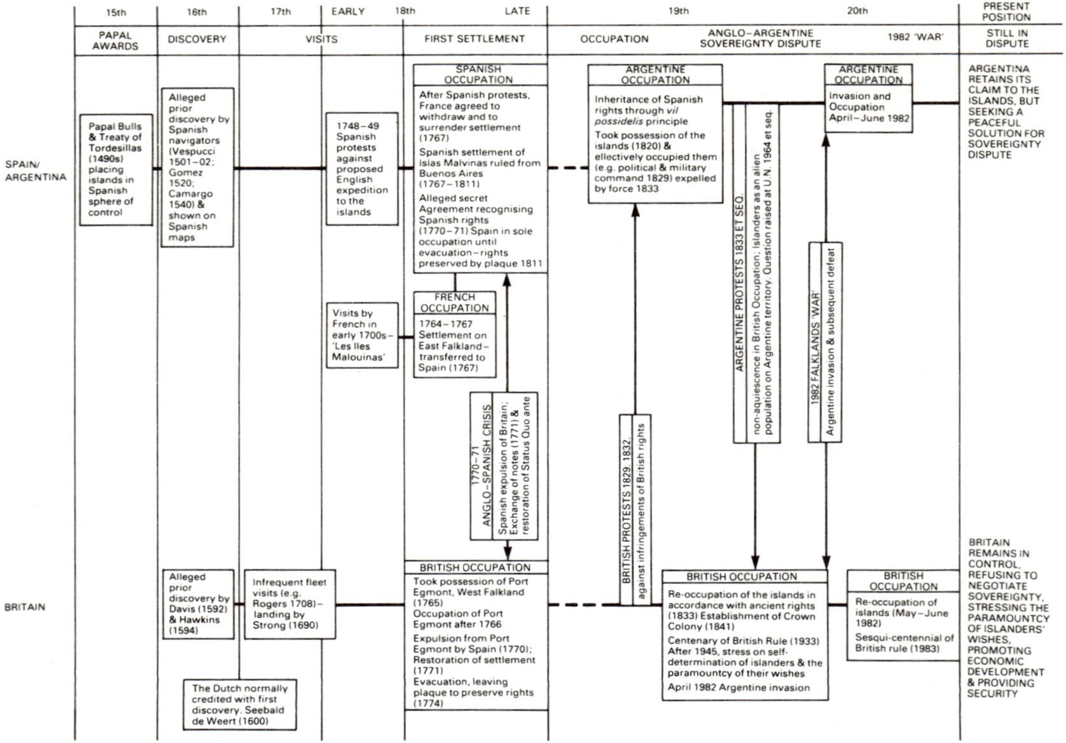

FIGURE 1.2 *Falklands or Malvinas?*

courage, and in resolution . . . We have ceased to be a nation in retreat.[24]

In fact, a recently published trilogy edited by Raphael Samuel was primarily inspired in reaction to such sentiments, and particularly by the nauseating wartime 'Gotcha-type' jingoism displayed by the British government, media and popular opinion:

> To anyone brought up on the war poets and associating naval expeditions with gun-boat diplomacy and the rhetoric of national glory with *1066 and All That* it seemed that the country had gone mad. The drowning of 500 Latin American sailors was treated as a national triumph, the sacrifice of British lives as a restoration of national greatness . . . Rather than treating this as an aberration we wanted to make sense of it in terms of the national past, and to account for the resilience of those sentiments on which history had seemed to turn its back i.e. to explain patriotism and jingoism through reasoned and historical enquiry.[25]

During the 1982 war the strength of the British claim to the Falklands was taken for granted by ministers, MPs, the media and the public, especially as governmental statements proved very effective in convincing the British people of the rights and wrongs of history; thus, Britain was right and Argentina, despite its protestations to the contrary, was wrong. Although there was an emphasis upon the right of self-defence and the need to uphold international law against an act of 'unprovoked aggression', the key part of the British rationale for the despatch of the Task Force centred upon the historical and legal 'facts' concerning the Falkland Islands. On 3 April 1982 Margaret Thatcher established the basic line that 'we have no doubt about our sovereignty' therein.[26] Soon afterwards a government publication entitled *The Falkland Islands. The Facts* – the very title a reassurance – made a strong appeal to the past in order to prove that 'British sovereignty over the Falkland Islands rests on a secure historical and legal foundation'.[27] A few weeks later another government publication, *The Disputed Islands*, reaffirmed that 'successive British governments have had no doubt about British sovereignty over the islands'.[28] The end of the war failed to remove this desire to press the British version of history; thus, the government's desire to establish the strength of the British case persisted into the postwar period, as evidenced by official submissions on the historical dimension to the Commons Foreign Affairs Committee as well as by the publication of brief Foreign Office 'official' histories in 1983 and 1986 stressing that 'the history of the period from 1494 to 1833 demonstrates the lack of foundation of the Argentine claim'.[29]

Inevitably, the official version of the past, in conjunction with the paucity of alternative British accounts, influenced the views of the media and opinion, which were encouraged to conclude that British sovereignty – to quote the *Daily Telegraph* – 'satisfies every known test of international law'.[30] Another so-called 'expert' asserted that 'those who have had occasion to conduct research into the history of this troubled little colony can only confirm that our claim to sovereignty is irrefutable'.[31] Within this functional context, Mary Cawkell, the author of two books on the territory, was presented by a former British diplomat active in the region's affairs as 'the historian of the Falkland Islands', whose 'passion for accuracy' was accompanied by a 'passionate belief in the Islands and their sturdy people'![32]

> The Argentine version of the history of the claim to sovereignty has gone round the world with the full resources of the Argentine Government behind it . . . How insignificant in practical terms the Argentine claim to sovereignty over the Islands was in 1960 may be judged from the fact that "Argentina" appeared only three times in the index [of her 1960 book]. This unhistorical statement encourages one to recall George Orwell's comment that 'the most powerful form of lie is that of omission'.

Rights accruing from alleged prior discovery of the islands by British navigators in 1592 and 1594, though often cited in certain accounts (for example Cawkell's), have tended to become less important for both historical and legal reasons. Problems of evidence render it impossible to prove that either John Davis or Sir Richard Hawkins actually discovered the Falklands, so that the facts of discovery, such as they are, do little for the British claim.[33] In any case, discovery confers a legal title *only* if followed by occupation within a reasonable period of time, and this condition was not satisfied by the British government.[34] Possession is often regarded as nine-tenths of the law, and inevitably the length, continuity, effectiveness and peaceful nature of its occupation became a key part of the British title deeds. In quantitative terms, it might be argued that over 160 years of British occupation (1766–74, 1833 to date) compares favourably with three years by France (1764–7), forty-four years by Spain (1767–1811) and *circa* eight to thirteen years by Argentina (1820–33). At present, Britain remains in effective and peaceful, albeit disputed, possession of the islands.

Occasionally, the British occupation has been presented, at least privately, in terms of the legal principle of prescription, according to which a title is acquired and strengthened through continued settlement. The distinction between 'occupation' and 'prescription' is somewhat blurred, but in October 1946 William Beckett, the Foreign Office's legal adviser, offered a revealing definition: 'Occupation is a method of acquiring title to territory which is *terra nullius* (ie. no-man's land), and prescription is a means of acquiring territory *which originally belonged to somebody else* and whose title is destroyed by the fact that the acquirer has successfully maintained possession'[35] From this perspective, Britain's occupation of 1833 might be interpreted – to quote from a 1946 Foreign Office memorandum – as 'an act of unjustifiable aggression which has now acquired the backing of the right of prescription'.[36] However, in 1983 the Foreign Affairs Committee reported that 'no legal argument in support of this prescriptive right has been formally advanced by the United Kingdom Government', perhaps because of the manner in which prescription suggests an originally weak, even illegal, title.[37] Nevertheless, there have been occasional references to the argument in Foreign Office memoranda, minutes and correspondence, as suggested in July 1982 by Francis Pym:

> The strength of our case depends on a detailed legal examination of all relevant events and factors. Even leaving aside arguments in our favour based on events before 1833, we have been consistently advised that our title can be soundly based on our possession of the islands from 1833. Our case rests on the facts, on prescription and on the principle of self-determination . . . Prescription as a mode of acquiring territory is generally recognised in international law.[38]

However, prescription raises a range of problems appertaining to both its general validity in international law and its specific relevance to the Falklands dispute. These points fall within the domain of international lawyers, but the relative paucity of case law means that there exists uncertainty concerning the extent of the time period (between twelve to a hundred years) required for 'acquisitive prescription'; whether the former owner's rights can be extinguished; whether an initially 'illegal' occupation can ever be validated; or whether continuing non-recognition and protest by the dispossessed party prevents the assumption of legal title.[39] Therefore, prescription offers uncertain guidance, quite apart from the fact that the Argentine government denies its relevance to the dispute: 'The British usurpation of 1833 cannot serve as the basis for a prescriptive title in favour of the United Kingdom, whose possession of the islands is illicit, disputed and not peaceful in nature'.[40] Over the years certain British diplomats have privately conceded that the occupation of 1833 was possibly 'illegal', at least according to present-day international law, although Professor James Fawcett, a former President of the European Court of Human Rights, has argued that title should be examined according to the principle of inter-temporal law, that is, 'in the light of the circumstances and the law applicable at the time'.[41] In any case, 'to say that force cannot give a good title is to divorce international law from the actual practice of nations in all known periods of history'.[42]

In general, Argentina has confined the sovereignty debate to pre-1833 criteria, that is, to the situation preceding the British occupation of that year. However, during recent decades the British government has attempted to move the discussion on to other factors, most notably towards an emphasis upon the 'paramountcy' of the islanders' freely-expressed wishes to remain British, which allegedly should 'take priority over any other interests involved', including the 'abstract' historical rights employed to cloak Argentine expansionism.[43] During April 1982 Margaret Thatcher, encouraged by the UN's emphasis upon 'the self-determination of peoples' and the need to take 'due account of the political aspirations of the people' (articles 1 and 73(e) of the UN Charter), asserted that 'we cannot allow the democratic rights of the Islanders to be denied by the territorial ambitions of Argentina'.[44] This approach was given renewed emphasis during the period following the 1982 war in order to substantiate the continued British commitment to the islands. 'It is highly relevant that the inhabitants of the Islands . . . have repeatedly made clear their wish that the Islands should remain British.'[45] The debate has been complicated by legal uncertainties relating not only to the definition of 'a people' but also to the validity of the principle. Self-determination, albeit often presented as a legal concept, is interpreted primarily as 'a political axiom . . . it has no strength in international law but great strength in international relations'.[46] In addition, Argentine accounts dispute the principle's applicability to the islanders, whose so-called 'rights' to choose their own future to 'Keep the Falklands British' stand in opposition to those of a dispossessed Argentine population. To some extent, this argument forced the British government to return to history in order to establish the 'scattered' and 'impermanent' nature of settlement on the islands prior to 1833. The self-determination argument remains 'part of our case to title', even if 'it is not frankly possible to quantify the extent to which our title rests on that particular element as opposed to others'.[47]

> The community on the Falkland Islands, though small, is a permanent, not transient, population . . . The Falkland Islanders have no less right to be accepted internationally as a 'people' with rights of self-determination than the population of Argentina.[48]

From this perspective, Argentina's invasion of 1982, far from ending British colonialism, would 'if allowed to persist . . . amount to colonialism in itself'.

The continuing Argentine challenge, in conjunction with changing international political and legal considerations, caused successive British governments to reappraise the nature and validity of their claim to the islands. This process gathered momentum during the twentieth century, when an escalating Argentine campaign focused attention upon the sovereignty issue. The extended closure and unavailability of relevant government files fails to prevent the reconstruction of a relatively complete picture of British official perceptions in order to indicate that the public image of certainty offered by British governments was qualified in private by a series of official doubts.[49]

Naturally, considerable attention was devoted to the precise legal status of the Falklands in 1833. At that time Britain presented its action as a reoccupation based upon pre-existing rights of discovery and settlement. Subsequently the British government, treating sovereignty as non-negotiable and conscious of its power advantage, devoted little thought to either the topic or the alternative version advanced in occasional Argentine protests.

During 1910 the Foreign Office produced a lengthy study of the nature and development of the dispute in response to a renewed Argentine challenge coinciding with the country's centenary celebrations. Henceforth, this memorandum, including its 1928 and 1946 revisions, became an accepted British point of reference.[50] This 'history' indicated various weaknesses in the traditional British case based upon pre-1833 criteria, since it cast doubt upon the 'facts' regarding initial British discovery, prior British settlement, the extent of British occupation before 1833, and the validity of alleged rights preserved by the 1774 plaque left upon the evacuation of the initial British settlement. Evidence relating to Argentine 'occupation' of the islands during the 1820s and early 1830s could be taken to imply that in 1833 Britain seized control of a legitimate Argentine possession. In practice, the memorandum cast the seeds of doubt and shook official confidence in the strength of a case simply taken for granted for a long time. The contents of the 1910 study were confined to a relatively small group of people even within the Foreign Office, and the British Embassy in Argentina did not receive a copy until 1927. In public, Britain's policy continued to be one of non-negotiation on sovereignty.

The debate conducted within official circles during the decades after 1910 about British title cannot be considered in depth here, but the series of discussions guided by the Foreign Office's legal advisers merely reinforced official doubts about the strength of the traditional British case on sovereignty.[51] It appeared that previous arguments centred upon prior discovery and settlement were relatively weak and inappropriate as compared to those emphasising the length and peaceful nature of British occupation. The 1933 centenary celebrations of British rule accentuated the force of the continuous possession argument. 'One hundred years' possession, whether disputed or not, should found a perfectly sound title to sovereignty in international law. Meanwhile, each year that passes, and in addition the celebrations of the centenary of Britain's occupation, strengthens His Majesty's Government's case.'[52] As a result, there occurred a major transformation in the British rationale for sovereignty. The new emphasis upon post-1833 rather than pre-1833 criteria was reaffirmed after the Second World War, and in September 1946 the Foreign Office codified the state of official thinking on the subject in a manner offering revealing insights of contemporary relevance.[53]

> The evidence for British priority of discovery is unsatisfactory, but irrelevant . . . The British case for priority of occupation . . . is untenable . . . [any rights preserved by] the leaden plaque of 1774 . . . lapsed . . . Great Britain has been in formal possession and effective occupation of the Falkland Islands since 1833: at that date the Islands were ineffectively occupied by the Buenos Aires Government, and it may be argued that they were thus open to acquisition by the first effectively occupying Power. In this sense, the British occupation of 1833 was, at the time, an act of unjustifiable aggression which has now acquired the backing of the rights of prescription.

The memorandum, though identifying the 'entirely British' nature of the population, made no reference to the principle of self-determination, which acquired ever-increasing significance during and after the 1960s.

Successive British governments have argued publicly that 'we are not in any doubt about our title to the Falkland Islands and we never have been'.[54] However, research establishes the existence of uncertainties and expressions of self-doubt, and particularly that the British rationale for sovereignty has been adjusted in the light of changing historical and legal circumstances. According to the Foreign Office:

> It would be misleading to attach any particular significance to what appear to be shifts in the factors which at any particular time carried most weight in the Government's thinking. Our case is based on all the facts and circumstances both before and after 1833. For example, the right of self-determination, which is now an important part of our case, has emerged only relatively recently as a principle of international law . . . With

> a subject as complicated as the history of the Falklands and the conclusions that may be drawn from that history for sovereignty, it is inevitable that there will be differing and, in some cases, conflicting views.[55]

The criteria employed to justify British sovereignty over the islands have displayed a reasonable degree of stability, even if the recent period has witnessed a downgrading of pre-1833 criteria, like discovery, alongside a growing emphasis upon the post-1833 occupation and the principle of self-determination. At present, Britain's allegedly 'strong' and 'very good case' to the islands rests upon the following rationale:

> The United Kingdom title to the Falkland Islands is derived from early settlement, reinforced by formal claims in the name of the Crown and completed by effective occupation for nearly 150 years. The exercise of sovereignty by the United Kingdom over the Islands has, furthermore, consistently been shown to accord with the freely expressed wishes of the people who form their permanent population.[56]

After 1982 British policy towards the dispute was centred upon Margaret Thatcher's blunt assertion that 'I am not negotiating the sovereignty of the Falkland Islands with anyone, they are British'.[57] The Prime Minister's visit to the islands in January 1983 symbolised her sense of commitment on the question, and British ministerial statements have reiterated that 'the sovereignty of the Falkland Islands is not available for discussion'.[58] Various reasons can be suggested as underlying this policy stance – a topic going beyond the scope of this study – but historical and legal factors provide no compelling reason for an unyielding British stand on the sovereignty question. Any objective study of the Falklands/Malvinas past seems likely to produce a 'grey' and uncertain result about title deeds in contrast to the traditional black-and-white version emanating from the British government. The historian can point also to the flexible attitude adopted towards sovereignty by British governments during the late 1960s, 1970s and early 1980s.[59] In addition, the focus of attention of recent British statements has shifted towards an emphasis upon the interests and/or wishes of the islanders, even if the nature of the Argentine counter-arguments to the principle of self-determination have forced the British government back on history.

Perhaps also for Britain, a country often criticised for 'national arrogance' in international affairs, the example of its superiority over a wayward Latin American country has proved a significant consideration. Indeed, the tone of Margaret Thatcher's previously quoted Cheltenham speech in July 1982 clearly recalled a Foreign Office comment made over fifty years earlier to the effect that 'it would be rather fatal to let people in South America think that they can treat us with impunity. This is really more important than the islands themselves'.[60]

Turning to the Malvinas past, the military invasion of April 1982 gave effect to Argentina's historical and legal rights to the Islas Malvinas, but the episode ended in defeat, the loss of the islands, and the overthrow of the military regime. Democratisation has fostered new lines of foreign policy, although Argentine governments whether headed by the Radicals or the Peronists in the persons of Raul Alfonsin (1983–9) and Carlos Menem (from 1989) respectively maintain the Malvinas claim as a policy priority. Like his predecessor, Menem disclaimed any intention of using force to secure the islands, and at the close of the 1980s this conciliatory stance provided the basis for an Anglo-Argentine rapprochement conducted under the 'sovereignty umbrella', which pushed that dispute aside in order to allow progress on other issues, including the resumption of diplomatic relations.[61]

However, Argentine governments and writers continue to identify the basic strength and foreign policy relevance of the country's claim. During November 1987 Dante Caputo, Alfonsin's foreign minister, reaffirmed the enduring power of history when informing the United Nations about 'the sound historical and legal arguments that support the Argentine claim'.[62] This dimension had been stressed a few months earlier when the great-great-grandson of Louis Vernet – the Argentine Political and Military Commander of the islands in June 1829 – informed the UN Committee of twenty-four about the continuing importance of 'the historical events of the Malvinas Islands'.[63] More recently, Carlos Menem employed his first UN address to remind the international community about his country's undoubted 'historical and inalienable rights to sovereignty over those South Atlantic islands'.[64]

The Argentine claim is 'founded on priority of discovery, priority of occupation, possession entered into and exercised, tacit and explicit recognition, and the acquisition of the titles belonging to Spain'.[65] Argentine accounts stress a series of rights inherited from Spain – these accrued mainly from discoveries made in the sixteenth century and settlement between the 1760s and 1811 – as well as Argentina's effective occupation of the islands prior to 1833. Inevitably, the British occupation of 1833 is depicted as resting upon force rather than law:

> In 1833, Great Britain, having no right on her side, could only resort to force in order to occupy the Islands . . . Force is still the cornerstone of Britain's presence . . . This act of force, this arbitrary and unilateral act was never and shall never be consented to by the Argentine Republic; and it cannot generate nor create any rights for Great Britain.[66]

As a result, the islands remain Argentine because it 'never consented to the United Kingdom's aggression and illegal occupation of the islands'.[67] The repetitive and emotive use of certain words and phrases like 'force' have characterised Argentine versions in vivid contrast to Britain's 'legal' interpretation stressing the implementation of 'ancient rights'.

Lacking power, the Argentine government was forced to rely upon diplomacy to press for the restitution of its rights: 'Since then, whenever possible, the Argentine republic has repeated its protests at the act of force and illegal occupation' and demanded the return of its islands'.[68] The resulting notes relied heavily upon history to expose the illegality of the British claim – this was presented as basically a function of British expansionism and imperialism rather than of legal rights – and to place on the record Argentina's 'indisputable rights' to the islands. Undaunted by British intransigence, Argentina exerted ever-increasing pressure upon Britain, notably, during and after the 1930s when Alfredo Palacios used parliamentary speeches, including a five-hour Senate speech on the subject in 1934, and writings to publicise the '*innegable*' (undeniable) nature of the country's rights.[69]

The changed international context of the post-1945 period fostered an alternative strategy. Argentina's search for wider political support encouraged an effort to present the 'facts' of its claim in a form designed to appeal to the growing Third World audience at the UN. This strategy, identifying the islanders as an alien, colonial population not entitled to self-determination, gave new life to the Argentine campaign, which was further encouraged by India's exploitation of anti-colonialism to justify the use of force against Portuguese Goa in December 1961.

> Colonialist policies . . . have an outstanding example in the case of the Malvinas Islands. At that time, advantage was taken of a country that was in the throes of organisation and struggling, as many new countries in Africa and Asia today, to achieve political and economic progress.[70]

The UN Committee of Twenty-Four on Decolonisation took up the question at the UN in 1964, when the Argentine delegation argued that the application of self-determination for the islanders would disrupt Argentina's national unity and territorial integrity.

> The Malvinas Islands are in a different situation from that of the classic colonial case. *De facto* and *de jure*, they belonged to the Argentine Republic in 1833 and were governed by Argentine authorities and occupied by Argentine settlers . . . they were evicted by violence . . . and replaced by a colonial administration and a population of British origin . . . The population is basically a temporary population . . . and cannot be used by the colonial power in order to apply the principle of self-determination.[71]

Paradoxically, the British reference to the self-determination principle, though intended to move beyond the rather sterile historical and legal debate, merely reinforced the Argentine resort to history in order to prove that alien 'colonists' implanted on Argentine territory were ineligible for coverage. This linkage of the issue with the UN demand for the termination of colonialism enabled the Galtieri regime to partially justify its military action in April 1982 on the grounds that colonialism was an 'act of force' and a state of 'permanent aggression'.[72]

The Argentine claim, though resting primarily upon historical, legal and decolonisation criteria, draws strength from other factors. Geopolitics has proved a significant preoccupation in Argentina, especially in military circles, where it has been utilised to establish that the Malvinas should be interpreted as part of one claim extending to Georgias del Sur, Sandwich del Sur and Antártida Argentina. Geopolitical thinking has reinforced historical accounts by advocating redress for large-scale territorial losses inflicted by Brazil, Chile, and Britain and pressing the geopolitical imperative for Argentina to control the Malvinas and other islands.[73] This spatial-power link and sense of encroachment upon the country's rights contributed to the 'indignant, self-righteous, combative, and hostile tone' characteristic of Argentine geopolitical writing.[74] Clearly, this dimension is relevant to any understanding of the Falklands/Malvinas problem, albeit an aspect exerting most influence in military circles. Jack Child has detected recent signs of moderation upon the part of Argentine geopoliticians in favour of a more integrative and cooperative approach, at least as far as South American countries are concerned.[75] During the 1980s emerging foreign policy trends, like Argentine-Brazilian cooperation, appeared to be in conformity with such thinking, even if its current influence upon Argentine policymakers remains debatable. Nevertheless, the impact of geopolitical thought should not be under-rated, if only because recent integrationist trends have often had an anti-British dimension.[76]

In many respects the Malvinas claim epitomises Argentine territorial nationalism, which has proved largely the product of geography and history teaching at schools and colleges as well as of a range of other educational/propagandist activities. The need for Ministry of Education approval for school textbooks permits political control over the content, and any survey of such books establishes the continued emphasis upon the Argentine version of the Malvinas past. In this manner, the teaching of geography and history implants, and then reinforces, the claim in the Argentine consciousness from a very early age. For example, *Manual Estrada*, a fourth-grade text, devotes considerable attention to the traditional historical explanation why '*las islas Malvinas nos pertenecen*' (the Malvinas islands belong to us); in fact, the justification for the restitution of '*nuestra soberania*' (our sovereignty) on 2 April 1982 was accompanied by an assertion that this action prompted '*una agresiva reacción de Gran Bretaña*' (an aggressive reaction by Great Britain).[77]

Similarly, history textbooks not only mention the dispute in detail – it is rare for any history textbook used in British schools even to mention the Falkland Islands – but also refer to the topic in terms of the '*usurpación*' of Argentine rights by Britain.[78] This process of 'education' continues in higher education, and opinion polls establish that those educated at university prove more sensitive to territorial issues than others, thereby highlighting the importance of what Carlos Escudé, an Argentine historian, has described as 'a self-perpetuating process of indoctrination through the educational system and the mass media'.[79] By law Argentine maps have been required to depict the islands as 'Las Islas Malvinas', while also using Spanish terms to describe the other areas claimed by Argentina, that is, South Georgia ('Islas Georgias del Sur'), the South Sandwich Islands ('Islas Sandwich del Sur') and Antarctica ('Antártida Argentina'). As a result, the maps used on postage stamps and in atlases, textbooks and other publications reaffirm the vast extent and apparent power of Argentina (Figure 3), while implying the British 'usurpation' of legitimate Argentine territory.[30] In addition, old maps dating back to the fifteenth and sixteenth centuries have been studied closely for evidence adjudged likely to reinforce Argentina's claim.

Inevitably, minimal effort has been made within Argentina to examine the country's version of the Malvinas past in a critical and balanced manner, such as to evaluate how far this past has been invented, even misinterpreted, for propaganda reasons in order to establish that 'the Malvinas have been, are, and will be Argentine'. A further problem relates to the extent to which the historical dimension constitutes a realistic rationale for current policy. A sense of commitment and emotion rather than of reason has determined thinking on the islands, thereby forcing writers to work within a narrow framework designed to establish the strength of '*nuestro derecho*'. The work of outsiders has been incorporated into Argentine writing, but only insofar as it reinforced existing viewpoints, as demonstrated by the selective citation of recent American and British research. Critiques are neither sought nor welcomed. Indeed, domestic political pressures create a difficult environment for unorthodox accounts, and naturally Argentine writers find it difficult to escape from the pressures regarding their country's superior historical rights to the islands.

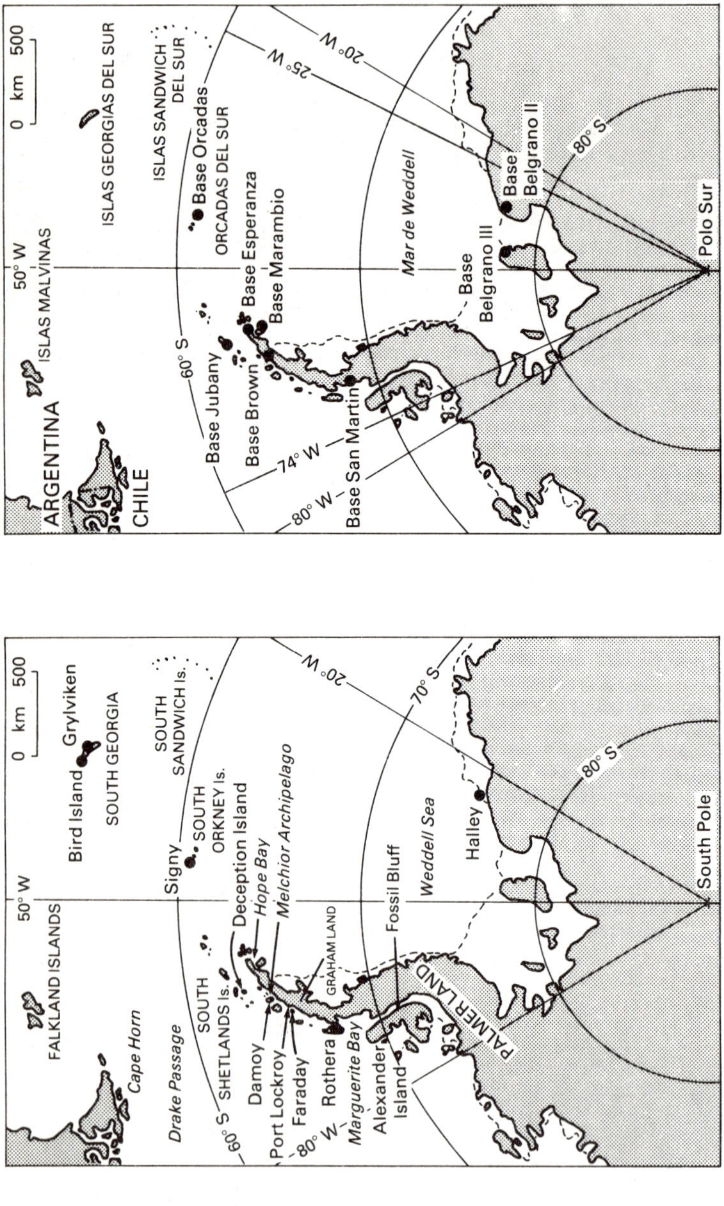

FIGURE 1.3 *Rival Anglo-Argentine perceptions of the South Atlantic region*

There exist only a few exceptions, mostly notably, the writings of Carlos Escudé. In 1985 the *Buenos Aires Herald* published a significant, even pioneering, series of articles by this Argentine historian, who exposed the invented and functional past in his search for the real historical 'facts'. He began by identifying the chauvinistic and self-perpetuating mythologies taken to be unquestionable truths in Argentina to the effect that 'we are (or should) be a great power' and that 'the Malvinas have been, are, and will be Argentine':

> These myths have perpetuated themselves, becoming deeply rooted in every stratum of Argentine society . . . and transmitted principally through the educational system. Thus, a vicious cycle is generated by which these untruths become more firmly entrenched, are taken for granted by rulers and ruled alike, and lead to the implementation of self-defeating policies.[81]

For each part of the claim, Escudé concluded that 'the matter is not obvious' – this phrase was often repeated in his series of articles – and hence the superiority of Argentine title is 'not something that can reasonably move anyone to indignation in 1985'.[82]

> The question of title deeds is much more complex and less obvious than Argentines generally believe . . . Who is right and who is wrong is not an obvious matter . . . I will not argue that Argentina has no rights, nor that her claim is weaker than the British one, but only that her rights are relative, stem from a highly complex analysis with many arguments and counter-arguments, that in the final instance it is very difficult to say objectively whose rights are better, and that it is therefore a case in which fanaticism is by no means warranted and much less so the recourse to force.[83]

More recently, Escudé analyzed the nationalist 'indoctrination' of Argentine public opinion through the educational process by reference to the manner in which school textbooks published since 1879 fostered perceptions of substantial territorial losses.[84]

> In order to get to the core of what is a contemporary problem in the sociology of knowledge which has deep political and economic reverberations, we must dig deep into history, with the objective of coming to some sort of conclusion with respect to whether the perception of territorial losses is or is not warranted . . . This perception is historically rooted.[85]

A cultural need to justify the absolute nature of Argentine territorial claims against both its South American neighbours and Britain explained the intense nature of the resulting historical and juridical debates and 'generated the need to lie'.[86]

Escudé has proved a relatively lone voice in Argentina – he is often dismissed as a maverick historian – and significantly his 1985 articles neither appeared nor attracted much interest in the Spanish langauge press. However, he influenced Villalobos' study in a provincial newspaper, and the resulting hostile reaction, including nationalist protests to the Ministry of Education, suggest one reason why Escudé has ploughed a lonely furrow within Argentina.[87] In the meantime, Argentina's claim, though possessing considerable merit in several respects, including 'effective occupation' in the pre-1833 period, is obscured by mythologies transmitted through the educational system and the media in the quest for certainty. Clearly time is required for re-education to break or disrupt the tyranny of the Malvinas past – even the events of 1982 are history now – in the light of the static nature of Argentine thinking towards a claim deeply embedded through all levels of society, as well as of the fact that the Argentine case is still pressed upon children in schools and colleges, as well as propagated throughout the country by the media and official statements.

The cause of international detente proved a central theme of Carlos Menem's speech to the UN General Assembly in September 1989. Regional conflicts threatened stability and required positive measures, as evidenced by his government's search for 'a peaceful solution' of the long-running Malvinas dispute.

> We have extended our hand, showing our willingness to enter into a dialogue while in no way casting any doubt on our historical and inalienable rights to sovereignty over those South Atlantic islands . . . The normalisation of bilateral dialogue must be followed by a firm and sensible will to bring

> an end to an obsolete colonial situation that is historically unsustainable ... We shall struggle unstintingly and peacefully, with reason and perseverance, to recover our Malvinas Islands.[88]

Argentina's projection of a more conciliatory image towards the dispute cannot disguise the reality that its position remains not only as inflexible as ever because of the enduring nature of the country's 'just claim' but also as ambiguous as ever because of the simultaneous emphasis upon 'struggle' and a 'peaceful solution'.

Defeat in June 1982 failed to terminate the campaign for the Malvinas. Argentina refused to go away and continued to challenge the legitimacy of the British position. Indeed, the events of 1982 gave substance, at least for a few weeks, to Argentine territorial aspirations and created images – these included television pictures of the Argentine flag flying over their islands and letters postmarked 'Islas Malvinas, Republica Argentina' – which survived the restoration of British control. Many Argentines have advanced the 'we will return' message – some refer to the 'first Malvinas War' – and the Alfonsin and Menem governments, though disavowing the use of force, have reaffirmed that the recovery of the Malvinas 'will be one of the central aims of our foreign policy'.[89] Indeed, during the 1989 Carlos Menem's presidential campaign was characterised by strong, even alarmist, oratory in support of the Malvinas claim. 'The pirates of the world should know that despite everything the Malvinas continue to be Argentine. We do not know how much time will pass or how much blood shall be shed, but our land shall return to the people.'[90] Such assertions might be dismissed as mere rhetoric intended for domestic political reasons – subsequently Menem claimed that he was only speaking metaphorically when mentioning the shedding of 'blood' – but the visitor to Argentina cannot fail to be impressed by the acceptance within all levels of society of the fact that the Malvinas belong by right to their country. Indeed, recent Argentine opinion polls, as recorded by Noguera and Willetts, have confirmed the broadly-based support for the claim, including an associated unwillingness either to surrender title or to accept a compromise solution; thus, the Malvinas do not constitute a divisive political issue between Peronists and Radicals.[91] Recently, certain Argentine politicians, including Dante Caputo (foreign minister, 1983–9) and José Bordon (Governor of Mendoza), have asserted that the sovereignty umbrella should never be interpreted as an indefinite arrangement given the eventual need to restore Argentine control over the islands.[92]

At present, the Argentine government remains imprisoned by the conviction that history is on its side, so that the claim is treated – to quote Oscar Camilion, a former foreign minister – as 'a very important priority . . . something that is beyond any kind of discussion'.[93] According to J.C. Metford:

> The tragedy of the present situation is that Argentina has created a myth which no rational or legal argument can destroy . . . The result is that every Argentinian is convinced of the justice of their cause and is prepared to die for it . . . For them ultimate sovereignty is a matter of national honour and no arbitrator will convince them that they should accept less . . . Two generations of Argentinians have been conditioned to an axiomatic irredentism which no factual discussion can remove.[94]

This was written at the time of the 1982 war, and should be viewed within that context. Democratisation has promoted new foreign policy perceptions, including a more sceptical approach to the past. Whether or not Argentinians are still 'prepared to die' for 'their cause' remains questionable, but the impact of history as a force for policy inertia and intransigence should not be discounted.

The extreme sensitivity, even obsessiveness, of Argentine governments and opinion to territorial matters imparts an added edge to any claim, while this dogmatic, myopic attitude qualifies Argentina's ability to conduct meaningful negotiations on the Malvinas with the British government. Opinion polls illustrate that Argentinians care significantly more about sovereignty to the islands than their counterparts in Britain; thus, it will prove difficult for any government to consider anything less than full and immediate sovereignty (leaseback, for instance) on account of the perceived political difficulty, if not impossibility, of selling any compromise solution to an electorate taught to believe in the strength of Argentine rights to the islands. This intransigent stance serves also to hinder the development of a rational policy based on perceived national interests and to further alienate the islanders and their British supporters.

The traditional stress upon 'how substantial our rights are' has hindered any engagement in a rational discussion regarding the real nature of Argentina's Malvinas interests.[95] The automatic, even unthinking, stress upon title deeds indicates a desire to recover a lost possession usurped by force. Repetition of this approach has tended to qualify other policy considerations, even if the impact of geopolitical and decolonisation considerations suggest the existence of both a strategic imperative to eject Britain in order to secure control of an Argentine sea and its resources, and an ideological desire to terminate an adjacent colonial presence. Successive Argentine governments, viewing current policy primarily as a logical extension of past events, have failed to clarify the country's real interests and priorities regarding the Malvinas. Policymakers and commentators seem content to propagate a series of platitudes relating to the importance of sovereignty without making any meaningful attempt to spell out national interests and priorities in the light of an ever-changing domestic and international context. The tendency has been to look backwards to the early nineteenth century and before rather than forward to the needs of Argentina in the 1990s and beyond, in a manner which encourages many to believe that the restoration of the *status quo ante* 1833 is all that matters.

Perhaps it is easy for outside commentators to over-exaggerate the historical and legal inertia characterising the attitudes of Argentine policymakers and opinion. How enduring are schoolday versions of the Malvinas past? What scope exists for pragmatism rather than dogmatism, especially given Menem's pragmatic approach? Opinion polls provide uncertain guidance on such matters, but most experts agree that a long time period is required before any significant progress becomes possible. 'Time, long time, must be allowed to do its work on the attitudes of all concerned.'[96]

Jeane Kirkpatrick has noted that the 1982 war had two acts: 'In act one, the Argentine occupied the Falkland Islands. In act two, the British reoccupied them. Act three has not yet been written.'[97] Of course, it is difficult to suggest how 'act three' will unfold, or even whether it will prove the final act of what has been a long-running saga. The 1982 war arose from an Argentine attempt to enforce its historical and legal rights in the face of perceived British intransigence. The resulting short, sharp conflict had a decisive outcome, the restoration of British control, but failed to resolve the fundamental sovereignty dispute. A difficult period of Anglo-Argentine relations, or rather non-relations, followed the war, and diplomatic relations were not restored until 1990, albeit only upon the basis of the so-called 'sovereignty umbrella', that is, a formula designed to shelve the central issue of dispute. Today, both disputants retain their rival claims and continue to attach significance to the public presentation of their respective versions of the past. Argentina and Britain, though displaying confidence in the rightness of their respective versions of the facts of history, agree to differ over the precise nature of the Falklands/Malvinas past.

Writing during the 1982 war, Bernard Crick emphasised 'the curse of sovereignty', when asking 'should we really worry whose flag flies over the islands rather than how the inhabitants are governed?'.[98] However, both disputants, imprisoned and divided by history, continue to worry about this 'meaningless' concept, which constitutes the root cause of the Falklands/Malvinas problem. Hence it is important to look backwards when going forwards into the 1990s and beyond in order to understand both the nature of the dispute and the influences upon current and future policy possibilities. Nevertheless, history has its limitations; thus, it proves difficult to provide a clear and objective answer to such questions as what is the Falklands/Malvinas past, which disputant possesses the better title to the islands, or whether 'historical aggression' justifies the contemporary use of force? As a result, the British government has attempted to move beyond this approach.

> While no doubt much time and energy could be spent in reviewing the history of the Falkland Islands between the first settlement in 1764 and 1833, and whilst the United Kingdom is confident of its legal case over the period, these facts cannot be allowed to over-ride the right of self-determination . . . It hardly seems appropriate to decide issues involving the welfare of people alive in the latter part of the 20th.century on the basis of (disputed) events in the early part of the 19th. century or even the 18th. century. If the international com-

munity were to discount 149 years of history, there would hardly be an international boundary which did not immediately become subject to dispute.[99]

Significantly, the continued British refusal to discuss sovereignty has been accompanied by an emphasis that the British commitment is to the islanders themselves rather than to territorial sovereignty. Thus, the 1989 Queen's Speech in Parliament reaffirmed that the British government 'will honour the commitment to *the people of the Falkland Islands*', as outlined a few weeks earlier: 'The Government remain committed to defend the Falkland Islanders' right to live in peace and security under a government of their own choosing'.[100] Margaret Thatcher's Christmas 1989 radio broadcast to the islanders attempted to reassure them in the light of recent Anglo-Argentine developments.

> You will have heard about our discussions with the new Argentine government . . . But right from the start we have made clear that we will not neogiate on sovereignty. That will remain our position and we shall uphold our commitments and responsibilities to you, the people of the islands.[101]

Similar sentiments were echoed by John Major, Thatcher's successor to the premiership, who employed his Christmas 1990 broadcast to alleviate the anxieties of islanders.[102] In the meantime, the tendency of both disputants to project their respective black and white versions of the Falklands/Malvinas past for policy purposes polarises, complicates and aggravates their relationship. Neither party is prepared publicly to admit that in reality claims are a grey and 'finely balanced' area, which is the view likely to be taken by any court in the unlikely event of the dispute being submitted for arbitration. Also it seems too much to expect historians ever to arrive at an agreed version of the Falklands/Malvinas past.[103]

One wonders, however, even after spending several thousand words on the subject, whether the historical and legal strength of either case really matters when it comes to the moment of truth. The events of 1833 and 1982 confirmed that, in the last resort, power and will constituted the decisive factors.[104] There seems a certain futility in relying too heavily upon historical and legal argumentation in sovereignty disputes – in such cases history and the law prove a function of policy considerations – when the prime focus should be placed upon the matter of advancing arguments which are politically reasonable, convincing and relevant to contemporary and future national interests. At present, history offers little hope for short-term changes in the Anglo-Argentine relationship, and there remains a need to 'confront these differing presentations of the past' in order to qualify the view that history often seems synonymous with propaganda.[105] We await still an in-depth, critical and balanced synthesis of the dispute's historical and legal dimensions, drawing upon Argentine, British and other research, and providing a more meaningful foundation for an improved Anglo-Argentine relationship.

The historical dimension explains in part why sovereignty has proved the master rather than the servant of events in the South Atlantic. During 1990 Sir Geoffrey Howe offered an illuminating commentary on the concept of sovereignty:

> Sovereignty is not some pre-defined absolute, but a flexible, adaptable organic notion that evolves and adjusts with circumstances . . . [and] constitutes a resource to be used, rather than a constraint that inhibits or limits our capacity for action.[106]

During November 1990 the conclusion of an Anglo-Argentine agreement on fisheries conservation in the South West Atlantic suggested that progress was possible on practical issues, even those touching upon disputed maritime boundaries.[107] However, the fluctuating fortunes of the negotiations caused by occasional legal difficulties indicates that a considerable amount of time will have to elapse before the concept of sovereignty becomes 'a resource to be used' in the Falklands question.

NOTES

1. Sir Anthony Parsons, British representative, 1 April 1982, quoted in *Britain and the Falklands Crisis: a Documentary Record* (London: HMSO, 1982), p. 24.

2. Raphael Perl (ed.), *The Falkland Islands Dispute in International Law and Politics: a Documentary Sourcebook* (New York: Oceana, 1983), p. 438.
3. Roberto Guyer, quoted in Michael Charlton, *The Little Platoon: Diplomacy and the Falklands Dispute* (Oxford: Blackwell, 1989), p. 103.
4. Daniel Kon, *Los Chicos de la Guerra* (London: NEL, 1983), pp. 46–7.
5. Laurio H. Destefani, *The Malvinas, the South Georgias and the South Sandwich Islands, the conflict with Britain* (Buenos Aires: Edipress, 1982), p. 5.
6. *Hansard Parliamentary Debates: House of Commons*, vol.21, col.25, 3 April 1982.
7. *Sunday Times*, 15 April 1990.
8. Michael Howard, 'Structure and process in history', *Times Literary Supplement*, 23–9 June 1989.
9. Eric Hobsbawm and Terence Ranger (eds), *The Invention of Tradition* (Cambridge: Cambridge University Press, 1983), p. 14; Peter J. Beck, 'Has history a future?', *Contemporary Review*, 249 (1986), pp. 264–7; Christopher Thorne, 'International Relations and the promptings of history', *Review of International Studies*, 9 (1983), pp. 123–31.
10. R.W. Davies, *Soviet History in the Gorbachev Revolution* (London: Macmillan, 1989), pp. 1–6.
11. Lech Walesa, *Path of Hope* (London: Collins, 1987), p. 36.
12. Davies, *Soviet History in the Gorbachev Revolution*, p. vii.
13. J.A.S. Grenville, *Europe Re-Shaped 1848–78* (London: Fontana, 1976), p. 349.
14. Arthur Marwick, *The Nature of History* (London: Macmillan, 1970), p. 11.
15. H. Tinker (ed.), *A Message from the Falklands* (London: Junction, 1982), pp. 188–90.
16. Michael Akehurst, *A Modern Introduction to International Law* (London: Allen and Unwin, 6th ed. 1987), pp. 15, 143.
17. See pro-Argentine advert in *New York Times*, 18 April 1982.
18. *Hansard (Commons)*, vol. 23, col. 25, 4 May 1982.
19. Letter by S. Miller to the author, 21 September 1987
20. For example, the UKFIC/Marplan Survey published 26 March 1986 recorded 94.5 per cent support for future British sovereignty. Argentine sovereignty secured 0.3 per cent.
21. Carlos Escudé, 'Argentine Territorial Nationalism', *Journal of Latin American Studies*, 20 (1988), p. 155; Harold Blakemore, *House of Commons Foreign Affairs Committee (HCFAC) 1982–3, The Falkland Islands, Report and Appendices*, p. 446.
22. *HCFAC 1982–83, Report*, p. xxiii.
23. *US mission to the United Nations, Press Release*, 22 May 1982.
24. *Conservative Central Office, Press Release*, 3 July 1982.
25. Raphael Samuel (ed.), *Patriotism: the Making and Unmaking of British National Identity* (London: Routledge, 1989), vol. 1, pp. x–xi; see also Anthony Barnett, *Iron Britannia* (London: Allison and Busby, 1982).
26. *Hansard (Commons)*, vol .21, col. 633, 3 April 1982.
27. *The Falkland Islands: The Facts* (London: HMSO, 1982), p. 3.
28. *The Disputed Islands* (London: HMSO, 1982), p. 29.
29. FCO, *Background Brief: The Falkland Islands – Early History* (London: FCO, 1983), p. 1; FCO, *Background Brief: Claims to the Falkland Islands* (London: FCO, 1986).

30. *Daily Telegraph*, 27 April 1982.
31. Letter from Sir Miles Clifford (former Governor of the Falkland Islands), *The Times*, 12 November 1982.
32. E.W. Hunter Christie, 'Foreword' in Mary Cawkell, *The Falkland Story 1592–1982* (Oswestry: Anthony Nelson, 1983). Sydney Miller has also praised Cawkell as a 'very accurate historian' whose work contrasts with the 'nonsense contributed by the Argentine government', S. Miller to the author, 22 March 1989.
33. Peter J. Beck, *The Falkland Islands as an International Problem* (London:: Routledge, 1988), pp. 35–8.
34. Ian Brownlie, *Principles of Public International Law* (Oxford: Clarendon Press, 3rd. ed. 1979), pp.149–50.
35. W.E. Beckett, Polar Committee Paper, 31 October 1946, quoted Beck, *The Falkland Islands as an International Problem,* p. 46.
36. Foreign Office memorandum, 17 September 1946, AS5782/311/2, CRS A1068/A47/26/6. Australian Archives, Canberra.
37. *HCFAC 1982–83, Report,* p. xxi.
38. Pym to Tam Dalyell, MP, 8 July 1982, in Tam Dalyell, *One Man's Falklands* (London: Woolf, 1982), pp. 137–8.
39. Ian Brownlie, *International Law and the Use of Force by States* (Oxford: Oxford University Press, 1963), p. 422; Akehurst, *International Law,* p. 295.
40. Perl, *Falkland Dispute,* p. 401.
41. *HCFAC 1982–83,* 17 January 1983, p. 140, p. 162.
42. John M. Lindsay, 'Conquest: a legal and historical analysis of the root of United Kingdom title in the Falkland Islands', *Texas International Law Journal,* 18 (1983), p. 130.
43. Cranley Onslow, *HCFAC 1982–83,* 21 February 1983.
44. *Hansard (Commons)*, vol. 21, col. 633, 3 April 1982; *HCFAC 1982–83,* 10 November 1982, p. 1.
45. FCO, *HCFAC 1982–83,* 10 Nov.1982, p. 1.
46. Fawcett and Sinclair, *HCFAC 1982–83,* 17 January 1983, pp. 142–3; Akehurst, International Law, pp. 296–297.
47. Sinclair, *HCFAC 1982–83,* 17 January 1983, p. 164.
48. *The Falkland Islands: The Facts,* p. 8.
49. Peter J. Beck, 'Research Problems in Studying Britain's Latin American Past: the case of the Falklands Dispute 1920–1950', *Bulletin of Latin American Research,* 2 (1983), pp. 5–13.
50. Memorandum by G. de Bernhardt, 7 December 1910, F0371/824/44753, Public Record Office, Kew, London; Lord Sherfield, *Hansard (Lords),* vol. 438, col. 187, 25 January 1983.
51. Beck, *The Falkland Islands as an International Problem,* pp. 49–55.
52. Memorandum by H. Goodwin, 3 February 1936, F0371/19763/A8083.
53. Beck, *The Falkland Islands as an International Problem,* pp. 53–5.
54. *The Disputed Islands,* p. 29.
55. Cranley Onslow, Minister of State at FCO, to the author, 17 September 1982.
56. FCO, 10 November 1982, *HCFAC 1982–83,* p. 1; *HCFAC 1982–83,* 17 January 1983, p. 168; Sir Ian Sinclair, FCO legal adviser, in Charlton, *Little Platoon,* pp. 3–6.

57. Margaret Thatcher, Christmas radio broadcast to the Falkland Islands, quoted in *Daily Telegraph*, 27 December 1984. For British policy since 1982: see Beck, *The Falkland Islands as an International Problem*, pp. 169–92; A.J. Groom, 'Britain and the South Atlantic: Politics and Strategy', paper at *Geopolitique et geostrategie dans l'hemisphere sud Colloque*, Universite de la Reunion, Reunion, May 1990.
58. Beck, *The Falkland Islands as an International Problem*, pp. 169–73; Lord Trefgarne, Minister of State for Defence Procurement, *Hansard (Lords)*, vol. 508, col. 1260, 13 June 1989.
59. Beck, *The Falkland Islands as an International Problem*, pp. 98–132.
60. Minute by R. Vansittart, 4 January 1926, F0371/10605/A6597; F.S. Northedge, *Descent from Power: British Foreign Policy 1945–73* (London: Allen and Unwin, 1974), p. 360.
61. Paragraph 2, *Joint Statement of Argentine and British Delegations* concluded in Madrid, 19 October 1989; *Joint Statement of Argentine and British Delegations* concluded in Madrid, 15 February 1990.
62. Caputo, 16 November 1987, Argentine government press release.
63. UNGA A/AC109/PV1327, pp. 31–3, 14 August 1987.
64. Menem, UN General Assembly, 25 September 1989, UNGA A44/PV5, p. 33.
65. Argentine government to Britain, 2 January 1888, quoted Beck, *The Falkland Islands as an International Problem*, p. 70.
66. Perl, *Falkland Dispute*, p. 363.
67. Perl, *Falkland Dispute*, p. 621.
68. Perl, *Falkland Dispute*, p. 364.
69. Alfredo L. Palacios, *Las Islas Malvinas: Archipelago Argentino* (Buenos Aires: Editorial Claridad, 1934), p. 127.
70. Perl, *Falkland Dispute*, pp. 364–5, p. 475.
71. Perl, *Falkland Dispute*, p. 368.
72. Perl, *Falkland Dispute*, pp. 478–9; Akehurst, *International Law*, p. 295.
73. Jack Child, *Geopolitics and Conflict in South America: Quarrels among Neighbours* (New York: Praeger, 1985), pp. 41–3.
74. Child, *Geopolitics in South America*, p. 43.
75. Jack Child, 'Antarctica and the Southern Ocean: Latin American interests' in A. Gauhar (ed.), *Third World Affairs 1988* (London: Third World Foundation, 1988), pp. 374–81; Virginia Gamba-Stonehouse, *Strategy in the Southern Oceans: A South American View* (London: Pinter, 1989), pp. 71–133; Jack Child, 'Latin Lebensraum: the geopolitics of Ibero-American Antarctica', *Applied Geography*, 10 (4) (1990), pp. 291–3.
76. Child, 'Latin Lebensraum'.
77. *Manual Estrada 4° Grado* (Buenos Aires: Angel Estrada, 3rd. ed. 1988), pp. 106–7. See also for younger children: Susana Molfino, *La Argentina: La Tierra, El Hombre, Sus Recursos* (Buenos Aires: Kapelusz, 2nd ed. 1985), p. 167. See also Peter J. Beck, 'Whose Island Story?', *History Today*, 39 (1989), pp. 8–11.
78. For example, see Jorge M. Ramallo, *Historia 3* (Buenos Aires: Ediciones Braga, 1988), pp. 11–13; Ernesto Palacios, *Historia de la Argentina 1515–1983* (Buenos Aires: Abelado-Perrot, 15th ed. 1988), pp. 316–17.
79. Escudé, 'Argentine Territorial Nationalism' pp. 156–60; Carlos Escudé, 'The Malvinas Conflict – 1', *Buenos Aires Herald*, 18 November 1985.

80. Peter J. Beck, 'Argentina's Philatelic Annexation of the Falkland Islands', *History Today*, 33 (2) (1983), pp. 39–44.
81. Escudé, 'The Malvinas Conflict – 1'.
82. Carlos Escudé, 'The Malvinas Conflict – III', *Buenos Aires Herald*, 25 November 1985.
83. Carlos Escudé, 'The Malvinas Conflict – IV', *Buenos Aires Herald*, 27 November 1985; Escudé , 'The Malvinas Conflict – 1'.
84. Escudé, 'Argentine Territorial Nationalism'.
85. Ibid., pp. 139–42.
86. Ibid., p. 159
87. Enrique Vera Villalobos, 'Las Islas Malvinas y La Acción Psicológica', *Rio Negro*, 1 August 1986; *La Nación*, 20 November 1986.
88. Menem, 25 September 1989, UNGA A44/PV5, pp. 32–3.
89. Alfonsin, 11 March 1983, quoted *HCFAC 1983–84*, Report, vol. 2, p. 149; Roberto Roth, *Después de Malvinas, Que?* (Buenos Aires: Ediciones La Campana, 1982), p. 12.
90. Menem, quoted *Somos*, 2 February 1989.
91. SOCMERC opinion poll, March 1990, *South Atlantic Council Press Release*, 26 March 1990; *Clarin*, 26 March 1990.
92. Dante Caputo, 'The new Anglo-Argentine dialogue' and José Bordon, 'The Future of Anglo-Argentine Relations', *EURAL/Royal Institute of International Affairs Conference on the Anglo-Argentine Dialogue*, Buenos Aires, 19–20 November 1990.
93. Quoted in Michael Charlton, *The Little Platoon*, p. 102.
94. J.C. Metford, in J. Goebel, *The Struggle for the Falkland Islands: a Study in Legal and Diplomatic History* (New Haven: Yale University Press, 1982 ed.), pp. vi–viii.
95. Destefani, *The Malvinas*, p. 5.
96. Sir Anthony Parsons, Comment on this paper, *The Falklands Conflict Conference*, University of Keele, September 1990, p. 2.
97. Jeane Kirkpatrick, 'The Falklands – a War the Americans lost', *Sunday Times*, 17 December 1989.
98. Bernard Crick, 'The Curse of Sovereignty', *New Statesman*, 14 May 1982.
99. Parsons to the UN, UNGA A37/582 Annex I, 28 April 1982.
100. Queen's Speech, *Hansard (Commons)*, vol. 162, col. 5, 21 November 1989; *Hansard (Commons)*, vol. 158, col. 435, 24 October 1989.
101. *Prime Minister's Christmas Message to the Falkland Islands*, 22 December 1989 (London: Prime Minister's Dept., 1989), p. 2.
102. *Prime Minister's Christmas Message to the Falkland Islands*, 12 December 1990 (London: Prime Minister's Dept., 1990)
103. *HCFAC 1982–83*, Report, p. xxiii.
104. Goebel, *Struggle for the Falklands*, p. 468
105. Marc Ferro, *The Use and Abuse of History* (London: Routledge and Kegan Paul, 1981), p. vii; Marwick, *The Nature of History*, p. 11
106. Geoffrey Howe 'Sovereignty and interdependence: Britain's place in the world'. *International Affairs*, 66 (1990), p. 676, p. 691.
107. *The Times*, 29 November 1990.

2 The Malvinas as a Factor in Argentine Politics

Peter Calvert[1]

The 'Falklands Factor' in the UK has a single, clearly defined meaning. It refers to the electoral advantage believed to have been given to Mrs Thatcher and to the Conservative Party by their response to the outbreak of war in the South Atlantic. There is no precise parallel to this in Argentina's more overtly nationalistic politics. The islands have meant different things to different groups at different times and there is no evidence that they have, as a specific issue, had significant electoral consequences at the national level. The purpose of this chapter is to identify the significance of the issue at various times before and since 1982. It is not its purpose to discuss the history of the claims and counterclaims to sovereignty over the South Atlantic islands. These are reviewed by Peter Beck in Chapter 1 and have previously been analysed both by him and by the present author elsewhere.[2] Here it is only necessary to recall that the origins of the Falklands/Malvinas issue, as seen by Argentina, by now have only a distant relation to the known historical facts. The assertion of sovereignty over the islands by Britain in 1833 certainly brought an immediate diplomatic protest from the government at Buenos Aires, but we can legitimately doubt if it remained for long a serious issue in Argentine politics, given that in 1838 the government of General Juan Manuel de Rosas opened negotiations to ask the United Kingdom to buy Argentina's interest in the islands, an offer which unfortunately the British government of the day turned down.[3]

The changing salience of the issue since the beginning of the twentieth century has depended at all times on the internal political situation in Argentina. Initially the emergence in 1916 of the Radicals as a mass party with broad electoral support did not alter this position. Though repeated formal protests had been lodged with the British Foreign Office throughout the nineteenth century, between 1888 and 1939 Argentine reservations about sovereignty were essentially confined to the marginal forum of the Universal Postal Union. In keeping with his attitude to other nationalist issues Yrigoyen's own administrations (1916–22, 1928–30) did not make much of it. His government fell to General Uriburu's coup in 1930 and Uriburu's government was succeeded by the military-backed government of General Justo and the Concordancia. The following year, in 1933, when the government of the Concordancia was engaged in the negotiations with Britain that were to lead to the Roca-Runciman Agreement, the coincidence of the centenary of the events of 1833 (commemorated in the Islands by a set of postage stamps showing historical maps of the region) led to further protests and in 1936 to the issue of rival stamps. Then in 1939, following the outbreak of war in Europe, Argentina formally recorded its claim to the islands as a reservation to the Declaration of Panama and restated it the following year at Havana, when Britain's endangered circumstances led Guatemala also to put in a bid for a share of Britain's colonial possessions in the event of its defeat.[4]

The Malvinas issue still did not really appear on the active political agenda in Argentina until well after the end of the Second World War. Even in the 1946 elections Perón's main campaign issues were domestic, until the publication of the US State Department 'Blue Book' offered him him the chance to denounce the opposition Democratic Union as a collection of quislings manipulated by the United States.[5] No mention of the question was made in Perón's first message to Congress or to the Armed Forces.[6] Only once he was in power did he cautiously and indirectly take up the issue of the Malvinas as part of his 'Third Position', and that by extending Argentine claims under the sector principle both to the Dependencies (whose sole association with the islands lay in their administration by Britain) and to 'Argentine Antarctica'.[7]

> We must mention the fact that we have sent a mission in the Antarctic. The Admiralty has collaborated efficiently with the Ministry of Foreign Affairs in the studies that took us to the declaration of Argentine rights on the continental platform and the Antarctic continent. Now that the expedition is back the Government hopes that very soon these inalienable rights will mean Argentine sovereignty over those lands that legally belong to her.[8]

There followed a short-lived crisis in diplomatic relations between Britain and Argentina. It is interesting how the crisis of 1947–8 foreshadowed later developments. When an Argentine announcement of a 'naval review' was followed by 'exercises' near the islands, the Attlee government responded by dispatching the cruiser HMS *Sheffield* from Simonstown to show the flag.[9] Subsequently Britain offered to submit the question of the Dependencies to the International Court of Justice, but this offer was turned down by Argentina. At the time, George I. Blanksten considered that few Argentines took their claim to the Dependencies and Antarctica very seriously.[10] Peronist propaganda however made much of driving 'the English pirates' from 'the national territory'. Such nationalist rhetoric was by no means new to Argentine politics, but enhanced political communications gave it a wider audience than had previously been the case, and the ability of the Peronists to mobilise vast crowds hinted at the degree of popular support it could command in the right circumstances.

The international context of Perón's initiative had originally been favourable. Argentina with its accumulated sterling balances was in a strong bargaining position, and Britain in the immediate aftermath of the war was on the verge of bankruptcy and willing to negotiate over British economic interests. However by 1948, with the onset of the Cold War and the Marshall Plan, Britain was recovering, while Perón's government was already coming under criticism abroad and had had to accommodate publicly to United States' leadership of the West. It does not appear moreover that the specific issue of the islands resulted in any significant increase in support for Perón himself. Despite the nationalist rhetoric and pride in *Argentinidad* which Perón fostered, and which contributed to the generalised climate of opinion which won him a second term in 1951, foreign affairs as such played little part in that victory. It is interesting that there is no specific reference to the question in the collection of extracts from Perón's speeches published in 1950, in which he rejects the charge of imperialism, or, rather, states that in Argentina it is an entirely constructive force.[11]

Domestic issues continued to dominate Argentine politics under Perón's immediate successors. Neither in 1958 nor in 1963 was foreign affairs a significant factor in determining the outcome of the elections. In 1960 Argentina supported UN

Resolution 1514 (XV) and by 1962 Argentine delegates were talking boldly of 'elimination of colonialism in all its manifestations and forms'.[12] Under the Illia government (1963–6) the Radicals obtained explicit recognition of Argentina's claims in UN Resolution 2065 (XX) of 16 December 1965, which defined the claim as an anticolonial issue. This resolution was not only a mandate to decolonise the islands by annexing them, but, even more conveniently, in keeping with the wishes of the anticolonial majority in the General Assembly, employed Resolution 1514 (XV) to set aside all previous historical claims as a basis for sovereignty. Though the resolution was not in fact binding in international law, it not only reinforced Argentine confidence in the rightness of its case, but also crucially weakened Britain's position in maintaining the right to self-determination of the island population. The irony of a country created by colonial settlement and by force claiming territory from another on the pretext of decolonisation seems never to have been understood in Argentina. Argentina is curiously blind to its own imperial history, and only in recent years have monuments to the Indians begun to appear in the territory won in the Desert War of 1879.

Yet the success of the Radicals in strengthening the Argentine position did not save Frondizi from military intervention in 1962. In fact an ill-timed visit to Antarctica on 8 March 1961 angered Chile without much affecting Britain, and Chile's reaction led the Argentine Congress indefinitely to postpone action on a protocol which would have solved outstanding boundary questions, including the future Beagle Channel islands dispute.[13] Nor did Resolution 2065 save Illia from the military coup led by General Onganía that toppled his government in 1966, and which was followed by a number of nationalist demonstrations aimed at the islands. Such demonstrations were associated with the Peronists, who thus sought to maintain their reputation as the more nationalist of Argentina's two main parties[14] and again revived the issue during their second disastrous period of office between 1973 and 1976.

Disputes over territory in Latin America have three causes: strategic, economic and symbolic.[15] The importance of symbolism is however so great that in many disputes it has outweighed all other considerations, and, given the dominance of the armed forces in Latin American politics, economics (in direct relation to the territorial dispute) has generally been subordinated to strategic considerations. In the case of the Falkland Islands there is no evidence that their economic value as such was of any interest to Argentina until United States speculation that their territorial waters might contain valuable oil reserves appeared to receive some confirmation from the Shackleton Report.[16] The effect of this was to strengthen the determination of the military government that took power in 1976 to resist any attempt to exploit what it saw as part of the Argentine patrimony. It should be said that this was a purely negative sentiment. The military government were not, as far as it can be ascertained, motivated in any way by commercial considerations or concerned about the context of trade relations with Britain or the European Community, from whose market they had effectively been excluded in 1975. For their part, the islanders did not realise (and still have not realised) that the greater the economic development of the islands, the more nationalist sentiment would be generated on the mainland.

And because geopolitics is currently unfashionable in Britain and Europe, British observers did not recognise (and still do not take seriously) the significance the Argentine armed forces attached to the islands' geographical position, which is said to have inspired Dr Henry Kissinger to describe them derisively as 'a dagger pointed at the heart of Antarctica'. This significance was far out of keeping with the intrinsic value of the islands, even on the most optimistic projections. Yet it should have come as no surprise to students of Latin American military thought. Argentina and Chile nearly came to war when in 1977 Chile was seen as seeking to expand into what geopoliticians call the arc of the Southern Antilles (*Arco Antilliano*), where the Argentine Navy had already established a clandestine toehold on Southern Thule in 1976.[17] Yet Chile even at this stage was seen in Buenos Aires as a diversion to the main military objective, the Malvinas. In military circles at least part of the islands' significance lay in their strategic position guarding the sea route from Chile to Buenos Aires. The capture of the islands would not only mean Argentina's dominance of this route, but the restoration of what was seen as part of the national patrimony, dominance of the *Arco Antilliano* and the elimination of Britain as a barrier to the extension of Argentina's South Atlantic sphere of influence.

By 1977 both Argentina and Chile were in the grip of extremely repressive military governments and Argentina was undergoing the throes of the Process of National Reorganization, intended to eliminate leftist revolutionary sentiment entirely. Hence there is, sadly, no way of measuring the popularity or otherwise of either issue at this stage, though later events show that there was little enthusiasm even in military circles for war with Chile. The mention of the Chilean navy has always had a sobering effect on army circles in Buenos Aires, though reference to Chile's army (unlike Brazil's) has been known to produce howls of laughter. Attention was diverted to Chile only after the crisis of 1977 was met by a firm UK response (see Chapter 6).[18] Once the Argentines were embroiled with Chile, their attention was preoccupied, and for the time being they were concerned to avoid a possible fight on two fronts at the same time. In the event the results of the Papal mediation, reluctantly agreed to in Buenos Aires at Christmas 1978, were not announced until 12 December 1980.[19] So the question of the Malvinas did not return to the military agenda until after the 'Dirty War' aspect of the Process was ended and economic difficulties multiplied. When Galtieri shunted aside Viola he needed allies to establish himself as President without relinquishing his key role as the army member of the junta. Galtieri (like Perón of Italian origin) was of recent immigrant stock and keen to score a nationalist triumph. He aligned himself with the naval hardliner Admiral Jorge Issac Anaya, the naval member of the junta and author of the so-called 'Goa plan'.[20] On the day of his appointment as Chief of Naval Operations, Vice Admiral Juan José Lombardo received instructions from Anaya to prepare a secret plan for the 'recovery' of the islands, but Argentines in general, who had treated the displacement of General Viola with absolute indifference, knew nothing of what was being prepared.[21]

The re-emergence of the Falklands/Malvinas issue was therefore intimately bound up with the nature of the military governments of 1976–82. The decision to invade was the last act in a long chain of events. It was not lightly taken but it was taken in haste and without due consideration. The military junta blundered into the adventure as a result of internal military politics. It was certainly not discussed outside military/governmental circles and the enthusiasm of the public was taken for granted. Those who planned it clearly believed that *when* successful their place in Argentine history would be secure forever. It would be an act of such massive nationalist significance that it would not be subject to the frequent reappraisals and critical re-interpretations that characterise Argentine historiography.

There was no shortage of plans of attack. Up to the last moment there were at least two in play simultaneously: an attack on South Georgia and the naval landing on the Falklands themselves. It is a measure of the politicised nature of the Argentine armed forces that the result in the end was an uneasy compromise, and with the knowledge now available to us from Buenos Aires it is clear that only a serious failure either of intelligence, or of decision-making, or of both, prevented the British government making an effective counter to the South Georgia incident and preempting the full-scale attack on the islands. The military government's previous record should certainly have left no doubt of its willingness to use force, abroad as well as at home. Apart from its confrontation with Chile, the first junta under Videla had already been active in intervention in neighbouring territories. Its role in the Bolivian coup of 1980 was well-known at the time and should have been within the knowledge of the British government. Between 1980 and 1982 Argentina had also been active in counterinsurgency operations both in the Southern Cone and (at the invitation of the United States) in Central America. Britain's failure to respond effectively to the South Georgia incident swept away caution. Those who had always argued that Britain secretly wanted to get rid of the islands won out and the landing went ahead, resulting in a huge surge of popularity for the regime. The impact of the recapture on the Argentine public was immediate – a frenzy of public rejoicing which made the military government momentarily appear popular.

> Even those commentators most used to the volatile and unpredictable nature of Argentine politics were amazed by the dramatic change in atmosphere that occurred within twenty-four hours. . . . The most poignant symbol of the change undoubtedly lay in the scenes of euphoria that pervaded the streets of Buenos Aires in response to the communiqué broadcast on national radio and TV by the junta. . . . In the eyes of many Argentines President Leopoldo Galtieri had

> indeed become the hero of a historic moment in which the country's armed forces recovered their *raison d'être*.[22]

> Some foreign commentators detected a hysterical note in the rejoicing. Roisin McAuley later wrote: 'The Argentina I encountered in a three-month stay during the Falklands War in 1982 was a fractured, neurotic nation – though Argentines kept telling me how the war had united the Spanish Argentines and the Italian Argentines and the German Argentines and so on, and had made them into one nation.'[23] But if it was tempered by anxiety the joy in a nationalist triumph was unalloyed by regret, for in both South Georgia and the islands themselves the landings were bloodless. The secrecy that surrounded Lombardo's planning from the outset was designed to ensure that the operation would be 'clean'.[24] The same troops who had taken part in the ruthless slaughter of their own countrymen took very untypical care to avoid loss of life among either the British defenders or the kelpers – in short they treated the occupation as a coup and not as an invasion. Argentines now see the invasion as a 'professional foul' and British reaction in actually embarking on a serious campaign to recapture the islands as disproportionate. At the time, the absence of casualties was used to justify the Church authorities, led by the military chaplains, presenting the war as fought for the defence of the Argentine nation and for Jesus Christ.[25]

Our ability to follow the course of public opinion during the crucial weeks of war is limited, though correspondence continued to get through with rather greater regularity than it does in peacetime. However there is no doubt that, owing to the restricted information available to the Argentine public, they were wholly taken aback to find that under the cover of censorship their government had been losing the war. For the censorship was powerfully reinforced by the self-imposed limitations of the press themselves, 'a mixture of unprofessionalism, cowardice and cynical opportunism' as Jimmy Burns puts it.

> In Britain there were Fleet Street newspapers like the *Daily Mirror* and the *Financial Times*, which, together with the British Broadcasting Corporation, adopted a balanced attitude towards the conflict – sometimes so "balanced" as to attract accusations from their rivals and government officials of be-

> ing too pro-Argentine. But in Argentina there was no such diversity of view. From beginning to end there was an eerie unison, in which the military was once again identified as the nation state.[26]

This phenomenon was to bring its own nemesis, as once the junta had embarked on a policy of suppressing the truth they seem increasingly to have come to believe their own propaganda. Having suppressed the news of the recapture of South Georgia, of the capture of Argentine prisoners at the Battle of Goose Green and of the success of the British landings at San Carlos Bay, they had left their people wholly unprepared for the news of the final surrender. Burns writes that,

> In the history books, 14 June 1982 went down as the day the Argentine nation experienced the full spectrum of emotion: from elation with the alleged heroic resistance to psychic breakdown at the realisation of humiliating surrender. Like a speeded-up film, the official communiqués came out one after the other, suggesting initially that the battle had only just begun and next that it was all over.[27]

Within hours the final realisation of the military failure had brought about a storm of protest against Galtieri who, unable to make himself heard by the crowd, was forced out without leaving an obvious successor.

Over the weeks that followed the military were in fact able to recoup their losses to a surprising extent. There was a strong military build-up under the interim government of General Bignone using the British creation of 'Fortress Falklands' as a pretext. Orders were placed for a nuclear submarine and aircraft, and losses of small arms and ammunition were made up. Some of this hardware was later cancelled by the Alfonsin government but as the military intended, the purchase of the new equipment coupled with the creation of the sort of refuelling capacity which the air force had lacked during the conflict, meant that the islands could be reached from the mainland. The same period saw the initiation of the Condor project with Egypt and Iraq. The latter power was also interested in acquiring missile capability. Condor 2, had it been successful, would have represented an even more radical tilt of advantage towards Argentina in the event of a future conflict. Yet in Argentina there was little or no interest in these developments – all public attention focussed on the newly realised fact of the Process and the return to democracy, and even abroad there was a general failure to realise how far military assumptions remained entrenched in Argentine territorial thought.

Perceptions of the issue since 1982 revolve round five main issues: the loss of the war; the military implications of the garrisoning of the islands; negotiations with Britain; the fisheries question; and relations with other countries in Europe and Latin America. It will be more convenient, however, simply to divide the period chronologically between the five-and-a-half years of Radical government and the period of the Peronist government which took power in 1989.

For the 1983 elections the Peronists made the fatal error of choosing as their standard bearer Italo Luder, a worthy workhorse with good links with the military but wholly lacking in charisma. His Radical opponent, Raul Alfonsin, used his own position as a lawyer who had defended victims of the Process to smear his opponent as soft on militarism, and he was successful in creating a broad but weak coalition which swept him to power with a huge majority. A major factor in his appeal was his insistence that under the new civilian government the armed forces would be brought under control.[28] At this point an encouraging signal from the British government might have enabled a lasting peace to be reached while the armed forces were still discredited, but that moment, unhappily, was allowed to pass.

From then on, it was Alfonsin's fateful decision to appoint an old friend, Dante Caputo, as Foreign Minister that ensured that the rapprochement between Argentina and Britain, so earnestly desired by perhaps a majority of Argentines, failed to take shape. Caputo took an activist view of his role, and was in fact later investigated for excessive expenditure claims in pursuit of his official duties abroad. However a wall slogan in December 1989 put the real problem pithily: 'The problem with Caputo is not that he travelled but that he came back'.[29] His strategy throughout the five years of Radical government was to try to put pressure on the British government through the United Nations and other international agencies to force them to concede a right not obtained by war; failing that he hoped that the 1987 election would return a Labour government which both he and Alfonsin believed as fellow social-democrats would be prepared to compromise. The Radical Party in 1985 even drew up a list of guarantees they would be prepared to give the islanders in such an event, though there was such an outcry from the Right when they leaked out that they were immediately disowned by the Foreign Ministry.[30] In fact, the diplomatic fiasco of the abortive meeting at Bern in 1984 had by then already put a stop to any serious possibility of resolving the issue on Radical terms. The meeting had been arranged through the good offices of the Brazilian government on the understanding that the question of sovereignty would be deferred until other more pressing questions were discussed. Within minutes the Argentine delegation had raised the forbidden question and the talks broke up.

The Alfonsin government started by seeing its domestic political position as stronger than it really was and ended by being unable to realise that its actual support had evaporated. From the beginning commentators noted the paradox of a centre-right coalition that was being led by a party dominated by a centre-left faction, which had secured its election only as the result of the defection of some 6 per cent of voters from the Peronists and 2 per cent from the Left. Core Radical support was never more than 25 per cent.[31] The peak of the government's success in internal politics came with the appointment of Juan Sourrouille as Finance Minister and the introduction of the Austral Plan in 1985. For a few brief months it seemed possible that Argentina had at last broken free from the inflationary spiral that had contributed to the destruction of military and civilian governments alike for a decade and a half. For the time being foreign affairs were irrelevant, and the question of the Malvinas almost forgotten, with the veterans of the campaign itself. As Felipe Noguera shows (Chapter 12), by April 1986 only 4 per cent of Argentine respondents still supported armed conflict over the islands and a significant majority (64 per cent) favoured re-establishing diplomatic ties with Britain.

In retrospect the Easter rising in April 1987 was the turning-point for the government. In the wake of the Papal visit the government attracted massive popular support for the defence of democracy in the face of a mutiny by nationalist junior officers. But peace was secured not on the basis of the strength of the government, but on its willingness to accept General Caridi's deal with the mutineers, many of whom were veterans of the Malvinas; a deal repeatedly denied by Alfonsin.[32] Once the principle of concessions had been established any serious possibility of reining in the armed forces was finished. Further concessions to the *carapintadas* followed the Villa Martelli uprising in December 1988, though Colonel Mohamed Ali Seineldin himself was posted to Panama, where he had previously helped train the army of General Manuel Noriega.[33] The Malvinas issue was taking a new shape as an issue in internal military politics, and it was one with which a civilian government was finding itself conspicuously ill-equipped to deal.

It is certainly significant that, despite the flood of controversy about the military government and the war, no one in Argentina questioned either the validity or the value of the Argentine claim to the islands. Only one scholar was prepared to tell his fellow-countrymen that the grounds for the claim were even slightly less than proved beyond doubt.[34] The Radical government made little or no effort to educate its citizens about the issue and an irritating feature of this period was the way in which every qualification or reservation about any aspect of British policy expressed by a British writer or speaker was rapidly hailed in Buenos Aires as 'proof' of the essential rightness of the Argentine position.

It was, on the other hand, sound policy to continue, and eventually successfully conclude, the negotiations with Chile over the Beagle Channel, for several reasons. The solution of this problem enabled the civilian government to display itself as one willing (unlike Britain) to accept international arbitration; it settled a long-running dispute with one of Argentina's neighbours; and left her free (in theory at least) to concentrate her efforts on the Malvinas situation. There were two problems with the Beagle Channel island issue which are relevant to question of the Malvinas. The first was that Argentina had no real grounds for its claim. The second was that even if it had been successful, there never had been any real glory in it. The dispute concerned the three uninhabited islets of Picton, Lennox and Nueva, in the eastern entrance to the Beagle Channel. The islets lay on the Chilean side of the deep water channel and the International Court of Justice (acting as the arbitrator) had no hesitation on the evidence in rejecting the Argentine contention that the real deep water channel lay on the other side and confirming that they were Chilean. Even the Argentine military government might have accepted this had President Pinochet of Chile not waited until they had done so before delivering himself of some rather foolish statements about the importance of the islands giving Chile access to the Atlantic (since Chile commands the entire southern shore of the Beagle Channel, well to the East of Cabo de Hornos, possession of the islets made virtually no difference to its strategic position). After two attempts at Papal mediation the issue was ultimately resolved in 1984 by some minor adjustments to the Vatican's original proposals which confirmed the rights of Chile. The real innovation of Alfonsin's government was to have the good sense to put the matter to a national referendum to avoid the Radicals being blamed for it.[35]

Resolving the dispute in this way, however, gave little satisfaction to the armed forces. It is important to differentiate between military attitudes to the war and military attitudes to the continued presence of Britain in the Malvinas. In military circles few blamed Britain for responding to a military attack in 1982; it was after all what they would have expected to do themselves. Even the sinking of the ARA *General Belgrano*, a cause célèbre at Westminster, was accepted as a regrettable consequence of war by Argentine admirals, including Admiral Anaya himself.[36] But the Argentine armed forces fully realised that the outcome of the war challenged the very reason for their own existence. It was significant that among the very few senior officers actually imprisoned for their part in the Process were the junta that lost the war. It was however quite unrealistic to imagine that for the foreseeable future a second attack on the islands would be successful, so for the time being the military seemed content to try to recapture their own internal political position.

In this context the continued presence of Britain on the 'national territory' continued to act as an irritant not only to the military but also to almost all civilians. Psychologists commented on the way in which Argentine schoolchildren took care to draw in the islands in their maps, while British children might omit large areas such as Devon and Cornwall.[37] It is likely that this merely reflects the systematic indoctrination on the issue that had become entrenched in Argentine education since

Perón – it is not clear that the schoolchildren were as careful in drawing in the Valdez Peninsula or sketching the outline of Misiones. Despite the general demilitarisation of Argentine education after 1983 a child learning to write in 1989 was still required to copy out '*Hoy es el 9 de Junio. Las Islas Malvinas son Argentinas*'; the teacher's comment, revealingly, was '*Por supuesto* (Of course!)'[38]

In Britain the Peronists had invariably been seen as the more belligerent of the two main parties on the issue, though as shown above this view is not wholly supported by the historical record, nor, in any case, were civilian views as significant as military ones. Sensationalist British press played up apparently belligerent statements made about the Malvinas by the Peronist presidential candidate, Carlos Saúl Menem. Given the location, Ushuaia, and the selective misquotation of a rhetorical question – 'how much time will pass, how much blood will be spilt, before the Malvinas become Argentine again'[39] – it was not easy to see at the time that in fact Menem's statements on the issue were the minimum to be expected in his position. In fact, the striking thing about Menem's campaign was that the Malvinas issue went almost without mention, and Menem himself at the time was quite definite that his policy with regard to the recovery of the islands had 'nothing to do with war. We have had a bad experience of that'.[40] In the event, the Peronists carried Tierra del Fuego in 1989 with 12,167 votes (42.1 per cent) to 10,428 (36.6 per cent) for the Radicals, who had carried the territory with a margin of only 0.2 per cent in 1983, and secured a resounding victory in the country at large, winning all but two of the twenty-four electoral districts to secure a decisive overall majority in the electoral college.[41]

It came as a surprise (not least to the Peronists) when Menem, after only a day in office, said he was ready to restore relations with Britain without discussing sovereignty in order to reach a *modus vivendi*.[42] With the issue of sovereignty 'under an umbrella' the two negotiators, Sr Lucio García del Solar for Argentina and Sir Crispin Tickell for the UK, reached a breakthrough in talks in New York between 16–18 August 1989, and announced an immediate restoration of consular ties in October.[43] Publicly the President defended the move as not abandoning in any way Argentina's historic position on the issue, while normalising its relations with the European Community

(EC) in order to focus its efforts on the major task of sorting out the economy. 'England cannot indefinitely hold sway over a land which she forcibly usurped in 1833,' he has said. 'Argentinians are not a bellicose people, but we demand social justice – which is the eventual return, through diplomatic means, of the Malvinas to Argentine sovereignty.'[44] However it was clear to strategists that, starved of funds by years of economic chaos, the Argentine navy, much of which had been put up for sale in 1985, was hardly able to put to sea, still less recapture the islands, and such forces as the much-reduced army could still train and equip were fully stretched patrolling the 5,000 km border with Chile.[45]

The question is, what factors other than economic, if any, lay behind the new government's change of policy on the Malvinas issue? It is not yet possible to answer this question with confidence. Questioned about it, Ambassador Garcia del Solar admitted freely that the decision to open negotiations with Britain came as a surprise to him personally. It was the personal decision of President Menem, he said; one of a series of decisions which he characterised as 'pragmatic', 'commonsense' and 'modern'.[46] In keeping with his style generally, it appears that the President wanted all outstanding problems solved at the same time. It is no secret that there was some opposition within the Partido Justicialista to the holding of talks, but by the time that they were concluded the endemic economic crisis so dominated matters that the argument that Argentina was sacrificing nothing of its position in order to obtain much-needed better relations with the European Community seems to have been accepted. In August 1990 the Peronist Menem, still with 56 per cent support in the public opinion polls, was warmly applauded at the opening of the 104th Palermo Show of the Sociedad Rural Argentina, when he criticised the political class and emphasised the importance of economic interests.[47]

The one section of Argentine society which did care deeply about the issue was that section which felt it had been forgotten, namely the veterans of the conflict. It was, ironically enough, the military government itself that had set the pattern, when in 1982 the repatriated captives were directed to land at Puerto Madryn in Chubut, Patagonia, in a vain attempt to avert the possibility of popular demonstrations against the government. By 1990 memorials had already been built to the war as far afield as Córdoba and Ushuaia, but only a small tablet recalled the war in the national capital itself. The inauguration of the Malvinas war memorial in Plaza San Martin revealed a bitter legacy of indifference and hostility to the veterans, who felt betrayed. Questioned about his feelings on the resumption of diplomatic relations with Britain the twenty-seven year old Eduardo Acosta, a former private soldier, said 'It's as if you let a man into your house and he raped your wife, then you invite him to dinner'.[48] Though this is a travesty of the historical record, unhappily it shows in the most forceful way that there can be no doubt about the real nature of the continuing importance of the question in Argentine politics. It may gain no votes, because every serious candidate must be nationalistic, foreign policy does not decide elections and the salient issues will be those concerned with the domestic economy. But it will not go away.

It is this that has led to the most striking paradox of Menem's presidency so far, that the man whom General Videla kept in prison for four years and under house arrest for a further eleven months has gone far beyond his predecessor in seeking to defuse the military question by amnestying the hundreds of military officers involved in the rebellions of 1987–8, and following this up by an amnesty for most if not all of those convicted of human rights abuses during the Process. When Alan Hunt, the British chargé, had arrived, a smart reporter (well versed, like many of his kind, in the art of putting his own 'spin' on questions) asked him on the steps of the Foreign Ministry for his views on the question of '*las Malvinas argentinas*'. In fluent Spanish he replied 'I am not able to answer a question like that' and went on to express his hopes for the success of the talks scheduled for February 1990.[49] February came and went; full diplomatic relations were restored and at the presentation of credentials the new Ambassador, Humphrey Maud, said that it was not the time to discuss Argentina's claim of sovereignty over the South Atlantic islands. For the first time for many years there was no significant dissent from this view.

Given this dramatic change in policy, some parallel or consequential changes in Argentine policy towards other countries could be expected, in particular in relations with Brazil and Chile. Friendship with Argentina's neighbours is undoubtedly a popular course, although public opinion is more concerned with economic integration in a world perceived as increasingly divided into trading blocs. In military circles Chile has long been regarded with the same suspicion with which their opposite numbers view Argentina, as Felipe Sanfuentes makes plain in this volume, and amongst civilians the belief that Chileans are infiltrating Patagonia, absurd as it is, was remarkably widely held under Alfonsin and no doubt still is. However, in the meantime, with the retirement of General Pinochet and the election of Patricio Aylwin the position in Chile has been changing, and Menem has shown his desire to support the civilian government in the clearest possible way, when he invited Aylwin to his inauguration while he was still only a candidate.

Relations with Brazil under the Radicals were 'very, very friendly',[50] and under Menem further impetus has been given to the creation of a unified market between the two countries, whose economies are in many respects complementary. President Collor is supposed to be Jorge Born's new protégé, and it was even said, following the collapse of his alliance with Menem in December 1989, that he might enter the Collor Cabinet. A close economic relationship with Brazil could have political advantages with the armed forces, helping to neutralise their perennial suspicion of Brazil's geopolitical ambitions. Anything that will strengthen Argentina's economy and enable it to pull back from the brink of bankruptcy will help the government that achieves it. However, Brazil's own economy under Sarney has hardly been a model for the rest of the world and it remains to be seen how effective the common market with Brazil will actually be in practice.

To sum up, therefore, the Malvinas issue has meant different things at different times, but common to all times and meanings has been that military, economic and symbolic issues are intertwined. For this reason it has not always been recognised outside Argentina that it is first and foremost a military issue. No 'deal' over the future of the islands has any chance of success unless the armed forces are able to accept it, and there is no solution that can be acceptable which does not involve the transfer of sovereignty. Unfortunately there has been and is a complete failure in Argentina to realise that in Britain too the question of the Malvinas could be a matter of honour. Given the general belief in Argentina that British 'intransigence' in 1982 and after was founded in the personality of Mrs Thatcher (a belief which Menem himself appears to share), her unexpected decision in November 1990 to resign and to pass on the torch to a newly elected Conservative leader inevitably raised new hopes in Buenos Aires.

But the paradox is that while successive Argentine governments have lost from their failure to handle the problem adequately, there is certainly no assurance that any Argentine government that gets the islands will secure any benefit as a result. The Malvinas issue is important in Argentina myth and culture but not in its electoral politics. Nationalism is like *machismo* – all politicians are expected to have it. No one notices if they do; everyone would notice if they did not.

NOTES

1. This chapter draws on research conducted during a visit to Argentina in December 1989 and January 1990, funded by the Nuffield Foundation.
2. Peter Beck, *The Falkland Islands as an International Problem* (London: Routledge, 1988); Peter Calvert, 'Sovereignty and the Falklands Crisis', *The Falkland Islands Journal*, 1987. For an excellent book-length discussion of the issues see Lowell S. Gustafson, *The Sovereignty Dispute over the Falkland (Malvinas) Islands* (New York: Oxford University Press, 1988).
3. Gustafson, p. 31.
4. The question is not discussed in Robert A. Potash's definitive work on the Argentine military, *The Army and Politics in Argentina, 1928–45; Yrigoyen to Perón* (Stanford: Stanford University Press, 1969).
5. Ruth and Leonard Greenup, *Revolution before Breakfast: Argentina 1941–1946* (Chapel Hill: University of North Carolina Press, 1947, reprinted Westport, Conn.: Greenwood Press, 1974), pp. 150–1.)
6. Coronel Juan Perón, *El pueblo ya sabe de qué se trata: discursos* (Buenos Aires, [1946], pp. 213–38.
7. Comisión Nacional del Antartico, *Soberania Argentina en la Antartida*, con nota preliminar del Presidente de la Nación, General Juan Petrón, 2a. edicion (Buenos Aires, Ministerio de Relaciones Exteriores y Culto, 1948).
8. Juan Domingo Perón, *Perón expounds his doctrine* (Buenos Aires, 1948, reprinted by AMS Press, New York, 1973), p. 241.
9. Peter Calvert, *The Falklands Crisis: the Rights and the Wrongs* (London: Pinter, 1982), p. 23.

10. George I. Blanksten, *Perón's Argentina* (Chicago: University of Chicago Press, 1974), pp. 431–2.
11. Juan Domingo Perón, *Perón parle. . . .* (Buenos Aires, 1950).
12. Gustafson, p. 63.
13. Alberto A. Conil Paz and Gustavo E. Ferrari, *Argentina's Foreign Policy: 1930–1962* (Notre Dame, Ind.: University of Notre Dame Press, 1966), pp. 200–1.
14. C.f. Susan and Peter Calvert, *Argentina: political culture and instability* (London: Macmillan; Pittsburgh: University of Pittsburgh Press, 1989).
15. Peter Calvert, *Boundary Disputes in Latin America,* (London: Institute for the Study of Conflict, 1983; Conflict Studies No. 146).
16. Gustafson, pp. 89–95.
17. Juan Enrique Gugliamelli et al., *El conflicto del Beagle,* (Buenos Aires; El Cid Editor, 1980), pp. 154 ff.; Beck, *International Problem,* pp. 117–9.
18. C.f. Michael Charlton, *The Little Platoon: Diplomacy and the Falklands Dispute* (Oxford: Blackwell, 1989), p. 64.
19. Juan Archibaldo Lanús, *De Chapultepec al Beagle: Politica Exterior Argentina: 1945–1980* (Buenos Aires; EMECE Editores, 1984), pp. 499–530.
20. Alejandro Dabat and Luis Lorenzano, *Argentina: the Malvinas and the end of military rule* (London; Verso, 1984), p. 92.
21. O.R. Cardoso, R. Kirschbaum and E. van der Kooy, *Malvinas: La trama secreta* (Buenos Aires; Sudamericana/Planeta, 1983), pp. 17 ff.
22. Jimmy Burns, *The land that lost its heroes; the Falklands, the post-war and Alfonsin* (London; Bloomsbury, 1987), p. 49.
23. 'The New Peron?', *The Irish Times,* Saturday, 24 June 1989.
24. Cardoso, Kirschbaum & Kooy, p. 20.
25. Burns, pp. 68–73.
26. Burns, p. 74.
27. Burns, p. 80; see also Latin American Newsletters, *Guerra de las Malvinas y del Atlántico Sur en el partes oficiales y comparativos* (London, Catálogos, 1983).
28. See Raúl Alfonsin, *La cuestión argentina* (Buenos Aires: Torres Agüero Editor, 1984).
29. Personal observation.
30. *The Guardian,* 19 July 1985.
31. Manuel Mora y Araujo, 'La naturalexa de la coalición alfonsinista', in Natalio R. Botana et al., *La Argentina electoral* (Buenos Aires; Editorial Sudamericana, 1985), pp. 89–107.
32. *The Guardian,* 14 December 1988.
33. *The Sunday Times,* 11 December 1988.
34. Carlos Escudé, *La Argentina vs. las Grandes Potencias; el precio del desafio* (Buenos Aires; Editorial de Belgrano, 1986); compare Alberto R. Coll, 'Philosophical and Legal Dimensions of the Use of Force in the Falklands War', in Alberto R. Coll and Anthony C. Arend, *The Falklands War: Lessons for Strategy, Diplomacy and International Law* (London: Allen and Unwin, 1985), pp. 39–44.
35. Hugo E. Alvárez Natale, *Beagle: de brujos y fantasmas a la decisión final* (Buenos Aires; Ediciones Politeia, 1984).
36. Charlton, *Little Platoon,* pp. 215–8.

37. *The Guardian,* 12 December 1985.
38. Personal observation.
39. *The Irish Times,* 16 May 1989; c.f. *The Sunday Times,* 9 April 1989; *The Daily Telegraph,* 15 May 1989.
40. Ibid.
41. Rosendo Fraga con Gabriela Malacrida, *Argentina en las Urnas, 1916–1989* (Buenos Aires; Centro de Estudios Unión para la Nueva Mayoría, 1989), p. 229.
42. *The Guardian,* 11 July 1989.
43. *The Guardian,* 19 August; 20 October 1989.
44. Ian Thompson, 'Menem: mystical messiah', *The Sunday Times,* 24 September 1989.
45. *The Sunday Times,* 25 August 1985; 21 May 1989.
46. Interview with Ambassador Garcia del Solar, Buenos Aires, January 1990.
47. *Buenos Aires Herald,* 12 August 1990; Michael Soltys, 'Situations vacant for opposition leader', ibid., 19 August 1990.
48. *Daily Telegraph,* 15 July 1990.
49. TV Canal Once, Buenos Aires, 7 p.m. news 21 December 1989.
50. Interview with Senator Gass, Chairman of the Senate Foreign Relations Committee, January 1988.

3 The Chilean Falklands Factor

Felipe Sanfuentes

Throughout the Falklands conflict, Chile played the role of a mysterious third participant whose presence was more felt than seen. During the war, the Chilean Falklands factor weighed more heavily in Argentine military strategy than is generally known. Conversely, within the British government, the Chilean connection was considered 'the most sensitive subject of the war'.[1] For these reasons, a good deal of speculation still remains about Chile's role in the Falklands War. However, the latter was only one expression of Argentina's territorial expansionism. Argentinian frustration over the failure to seize the Chilean Beagle Channel Islands was an important element in the decision to invade the British Falkland Islands and their Dependencies. Like the Beagle Channel dispute, this aggression was part of a premeditated plan to secure foreign strategic outposts in the South Atlantic and South Pacific areas. Territorial claims were disguised under a game theory of sovereignty whereby any claimed territory would become an imaginary 'Argentinian sovereign territory'. The Falklands adventure led the Argentine dictatorship to develop a two-front military strategy out of fear of a Chilean military attack – despite Chile's hasty and unnecessary proclamation of neutrality in the conflict.

The territories that comprised the Viceroyalty of the Rio de la Plata never belonged either to the United Provinces of the Rio de la Plata or to present-day Argentina, which was founded as late as 1862. It is therefore obvious that Bolivia, Paraguay, Uruguay and Misiones did not segregate themselves from the United Provinces of the Rio de la Plata, but from the Viceroyalty, that is, not from Argentina, but from Spain.[2] The city of Buenos Aires became the Federal Capital of the new Argentina only in 1880.[3] Geopoliticians like General Osiris G. Villegas, have tried to explain what they consider to be Argentina's territorial 'losses' by comparing two maps of South America, one of 1776 and one of 1978. The map of 1776 showed present-day Argentina, plus Bolivia, Paraguay, Uruguay and Chile south of the Bio Bio River. The 1978 map showed, however, Argentina minus Bolivia, Paraguay, Uruguay, and other territories 'lost' to Brazil and Chile.[4] It might have been equally revealing if a third map had been included showing the Argentina of 1833 which possessed neither Patagonia nor half of Tierra del Fuego. It was only in 1881, after the signature of a treaty, that Argentina gained its legal titles over Patagonia, half of Tierra del Fuego and the Isla de los Estados. Chile received the Straits of Magellan, half of Tierra del Fuego, and *all* the islands south of the Beagle Channel.[5]

The Province State of Buenos Aires and the United Provinces of the Rio de la Plata had as their southern boundary the Diamante and Quinto Rivers in the west, and the Colorado River in the east.[6] South of that boundary line was the Reyno de Chile, which had as its eastern frontier the Atlantic Ocean, according to the 1775 map of Spanish cartographer Juan de la Cruz Cano y Olmedilla.[7]

Argentina's irredentist writers used as a dominant theme the notion that Chile is an expansionist country. Concerning this 'expansionism,' the Chilean historian Francisco Encina has noted that Chile participated in 1820 in the expedition to liberate Peru without claiming compensation: it was motivated solely from the desire to see Peru independent. Secondly, in 1835 during Argentina's period of anarchy, the provinces of Cuyo (Mendoza, San Luis and San Juan) asked to be reincorporated into Chile, in a letter addressed to Minister Diego Portales, but he ignored the request. Thirdly, Chile did not seek any territorial or economic compensations after its victorious war against the Peruvian-Bolivian Confederation (1839). Fourthly, after Chile's military victories of 1879–81, in the War of the Pacific against Peru and Bolivia, it took, as compensation, the provinces of Antofagasta and Tarapacá in which 80 per cent of the population was Chilean. Lastly, after the War of the Pacific, Chile had the strongest armed forces of Spanish America, and yet it renounced its titles to the 700 000 square kilometres of eastern Patagonia for the sake of peacefully settling its dispute with Argentina.[8]

Under the Treaty of 1881, Chile even went to the lengths of agreeing to the permanent demilitarisation of the Straits of Magellan. In return for this unnecessary concession, Argentina eventually militarised the neighbouring territory of Tierra del Fuego, and today has a network of military and air bases north and south of the Straits of Magellan.

The United Provinces of the Rio de la Plata and the State Province of Buenos Aires expanded to become the eighth largest country in the world. This was after conquering the Paraguayan province of Formosa with the help of Brazil and Uruguay, and later acquiring additional territories through arbitration.

Argentina's objective in the Beagle Channel area has always been the control of the southern passes that link the Pacific and Atlantic Oceans and lead to Antarctica. To help itself in this scheme, a geopolitical slogan was devised, 'Argentina in the Atlantic, Chile in the Pacific'. Article 2 of the Protocol of 1893 states clearly that Argentina cannot claim positions on the Pacific coast nor Chile on the Atlantic coast; nowhere is mention made of insular territories or territorial seas. Through Papal mediation, Argentina finally succeeded in intruding in the Pacific Ocean, but without securing a single islet in the area.

In Chile, some boasted that their country did not have to yield one inch of land through the Papal mediation process. Regardless, Chile did renounce its substantial maritime rights, and accepted some convoluted maritime servitudes, such as granting 'innocent passage' to Argentine warships through internal waterways between the demilitarised Straits of Magellan and the Ushuaia naval base on the Beagle Channel. A very similar Argentine proposal for innocent passage through Chilean internal canals had been overwhelmingly rejected by a Chilean democratic congress in 1960.[9] With such a precedent, it is a mystery how a Chilean military government could have found the same burdens supportable in 1984. Through the Treaty of Peace and Friendship, Chile gained recognition of sovereignty over islands it had occupied for over 100 years.

The Argentinian approach to sovereignty issues seemed to have been inspired by what could be called a special game theory of strategy, far removed from the mathematical game theory of strategy. As is well known, territorial sovereignty signifies that a state has total control over its territory, which would include enforcing laws, executing decrees, and exercising overall jurisdictional power inside its boundaries.[10] From this point of view, *none* of the territories claimed by Argentina could have been or could be considered to be under its sovereignty. Of sovereignty in general, Michael Akehurst has rightly observed that 'it is doubtful whether any single word has ever caused so much intellectual confusion and international lawlessness'.[11]

Argentina threatened to invade Chilean sovereign territories on the Beagle Channel and then invaded British sovereign territories in the Falkland Islands and their Dependencies. Argentina resorted to a demagogic argument justifying seizure of those territories, by claiming that there was no such thing as an invasion when 'its' territories were being 'usurped' by another power. The strategy followed by the Argentine military government had two variants: to occupy territories in order to negotiate later (*ocupar para negociar*) or to negotiate in order to occupy later (*negociar para ocupar*). The single objective was the occupation (annexation) of foreign territories. With this one-track objective in mind, Argentina in fact had no room for negotiation or compromise, although these territories were not under its sovereignty. From this point of view, if after seventeen years of negotiations the British had 'failed' to transfer to Argentina *their sovereignty* over the Falklands, then the invasion of the Islands was 'justified' – indeed, provoked. Thus the Beagle Channel invasion plan and the Falklands invasion plan were closely linked from the point of view of Argentine expansionist geostrategy. Both of these conflicts were fabrications to justify Argentinian claims against Chilean and British insular territories in the South Pacific and South Atlantic Oceans.

An interesting case for the study of Argentine politico-military stratagems was provided by the Beagle Channel quasi-war of December 1978. There are two possible hypotheses about Argentina's military intentions on this occasion: it is impossible to be certain whether Chile faced a serious threat, or only an elaborate deception.

In January 1978, the Argentine government declared the Beagle Channel International Arbitration Award incurably void (*insanablemente nulo*). In Chile, this decision had the impact of an earthquake and was described as 'juridical savagery'.[12] Whereas the Treaty of 1881 explicitly recognised the Beagle Channel as Chilean (under article 3), the Beagle Channel Arbitration Award moved the boundary line to the Channel's median line. By this decision, Chile was not necessarily a winner in the litigation. However, in Chile, government and media interpreted the award as a victory because Chile's sovereign rights over Picton, Lennox and Nueva Islands in the Beagle Channel had been recognised. For her part, Argentina's clear progress did not satisfy her expansionist ambitions. In their search for justifying Argentina's unlawful action, Argentine geopoliticians went as far as exposing 'Anglo-Chilean complicity' in the arbitration award. These geopoliticians claimed that because Chile and Britain have mutual strategic interests in the area, the British Crown could not be considered 'independent'.[13] Needless to say, the British sovereign played only a symbolic role, for the Court of Arbitration was composed of five judges, of different nationalities, who were all members of the International Court of Justice in The Hague.

After declaring the arbitration award 'null and void', Argentina mobilised its armed forces and conducted large-scale military maneouvres with the ostensible purpose of 'recovering all the Argentine South Atlantic islands' that were occupied 'illegally' by Chile. Simultaneously, Argentina went on a military equipment purchasing rampage of which its economy is still suffering the consequences today. In December 1978, signs were posted in Ushuaia – among other places – which proclaimed: 'Picton, Lennox and Nueva are Argentine'.[14] In Buenos Aires and other Argentine cities, air defence exercises took place. This carnival-like military activity, together with a strident anti-Chilean propaganda campaign, spread panic among the Argentine population, but did not intimidate their Chilean neighbours. The Chilean armed forces were, however, in a high state of readiness, and were quite able to take care of any external military aggression. Colonel Sir Thomas H. Holdich's observations about Chile's military strength, on the occasion of King Edward VII's Arbitration Award of 1902, were probably still pertinent in 1978:

> On paper the Chilean armed military strength may appear insignificant, but the latent strength of the country is not to be measured in figures. Its foundations are laid in the sturdy spirit of patriotism which has struck its roots deep into the heart of the people.[15]

In 1987 the Argentine magazine *Somos* published two articles about how the war against Chile had been planned. In a game of sovereignty, it was fitting to have called this quasi-war *Operativo Soberania* (Operation Sovereignty). The invasion was set to commence on 22 December 1978 at 2200 hours. The first phase of the operation would have consisted of the occupation of Picton, Lennox and Nueva Islands in the Beagle Channel, plus several other islands, including Cape Horn Island. In this phase no 'serious' Chilean resistance was expected. The city of Punta Arenas would have been bombed, as would its lines of communication and supply. On 23 December at 2400 hours, the army would have advanced into Chilean territory at four different points. The city of Santiago would have been the first great objective, the port of Valparaiso a possible alternative. The Argentine offensive would not have been limited to the southern sector. A second phase of the invasion would have been launched from Neuquèn Province, thus cutting Chile in half. The air force would have made massive attacks on all Chilean airports in order to destroy as many Chilean airplanes as possible. Air and naval battles in the Magellan and Beagle Channel areas would have been expected to take place. It was feared that Chilean armoured fighting vehicles could have advanced towards Comodoro Rivadavia through Chubut Province's mountain passes where it was thought that Argentina had its weakest defences. Forty thousand soldiers were concentrated in Comodoro Rivadavia for the defence of the city. According to *Somos*, the Chileans would have had three different choices of attack: the city of Rio Gallegos, Neuquèn Province, and through the north, without specifying where. In 1978, the Argentine army was considered larger than the Chilean army in a proportion of 1.8 to 1 (in 1990 the reverse was true). The air forces and navies of both countries were considered on a par, with a slight Argentine advantage. The objective would have been to reach a quick victory based upon the destruction of the Chilean military machine. The most optimistic forecast anticipated a Chilean surrender; another one expected Chilean acceptance of Argentine territorial claims. Military operations would have been followed by a withdrawal of Argentine forces towards the border.

Argentine planners also foresaw the possibility of an expansion of the conflict with Peruvian and Bolivian intervention in the north. It was thought that they might have been tempted to try to recover territories lost in the War of the Pacific. The intervention of an international organisation, presumably the United Nations, was expected as a possibility to stop the war, in which case Argentina would have immediately accepted the presence of a peacekeeping force. Military operations had been planned in such a way that there would have been a short and decisive attack, which would have conquered a maximum of territory in a minimum of time. The Argentine armed forces joint command staff estimated that both sides would have suffered about 20 000 casualties in an armed conflict that all Argentine planning had conveniently predicted to be a short 'victorious' war.[16] At the time, there had been rumours that Argentinian hardline generals were so confident of a military victory that they had boasted to their officers that soon they would be 'pissing in the Pacific'![17]

Vatican intervention stopped *Operativo Soberania* only hours before its scheduled commencement. As D-Day was nearing, Argentina realised that Chile had promptly fortified and reinforced each and every island that was to be invaded. On 21 December, Argentinian UN Ambassador Enrique Ros sent an urgent note to the Security Council complaining that Chile had fortified Picton, Lennox, Nueva, Freycinet, Wollaston, Horn, Herschell and Deceit Islands, and that it thus was altering the *status quo* of the area. He also indicated that Chile had taken these 'illegal' measures in order to create a *fait accompli.* Argentina had not done anything similar out of 'respect' for the mediation process.[18] These timely Chilean countermeasures saved the Chilean Beagle Channel Islands from a premeditated Argentine *coup de main.*

On 22 December, the Chilean fleet steamed towards the area where the Argentinian navy would be giving support to the occupation of the Chilean islands. Only hours before the deadline, the Chilean ships received orders *en clair* to counterattack. There was also a Chilean submarine poised to attack the aircraft carrier *Veinticinco de Mayo.*[19] General Pinochet had warned that Chile could place half a million men under arms in case of war.[20] Pinochet had also made it clear to the Argentine negotiators that an attack against any Chilean position, north or south of Chile, would start a general war. This was contrary to what the international press had been led to believe – that Argentina would just deliver a *coup de main* in the Beagle Channel, and follow through with negotiations.

Unlike Argentina, Chile did not go through a period of war hysteria. General Nilo Floody, the Chilean army commander-in-chief of the Magellan military region, declared the morale of the troops to be excellent and stated that they clearly knew why they were defending Chilean sovereignty in the area. He also commented that the Chilean soldier did not need artistic shows or musicians to raise his morale. He was referring to processions of musicians and show people which were sent to raise the morale of Argentine troops stationed on the other side of the border. 'When a soldier needs entertainment to boost his morale, it is because something is failing.'[21] In Punta Arenas, a 'hot' zone, situated in the Magellan strategic area, invasion was perceived as certain in December 1978. For this reason, there is still today among the inhabitants a much deeper degree of distrust towards Argentina than in other parts of Chile.

The Argentine Admirals Emilio Massera and Jorge Isaac Anaya were extreme hardliners in time of peace, but when war became a distinct possibility in the Falklands, as in the Beagle Channel, they quickly withdrew the fleet to safe havens. On 1 July 1982, Brigadier Ernesto Horacio Crespo, Commander of the Argentine Air Force South, addressed an interesting secret memorandum to the Argentine Air Force Chief of General Staff about the navy's perplexing inactivity during the Falklands War, asserting, among other things, that

> the naval force was dedicated to fighting for national political ends and to conserving its own forces, while devoting a minimum effort to the war. This fact, which repeated a situation similar to the 1979 Beagle Channel crisis, causes one to have doubts about its real intentions or professional capacity in the defence of the nation. Everything expressed above becomes even more important when analysing the Falklands conflict which was entered as part of the navy's permanent planning and motivation. Although not having a defined war plan, this same force was assigned command of the theatre of operations, but it had neither the capacity nor the will to assume it, despite its own declarations. In conclusion, the navy has no fighting spirit and it is unreliable in its actions, and it faces only those operations in which it has direct interests.[22]

Nonetheless, the lack of fighting spirit was not limited to the navy. A large portion of Argentine military personnel was composed of conscripts. The conscripts who would have fought the Chileans were the same as those who were later sent to the Falklands. Concerning their equipment, the *Gaceta Argentina* reported that Argentine conscripts were as well equipped as British soldiers. As their own propaganda had it,

> The British are armed with FAL rifles identical to ours. MAG machine-guns, identical to ours. But they don't have 120mm mortars [nor 155mm artillery].
>
> The Difference Between Us and the Enemy –
> The enemy does not know the reason why he is fighting.
> The enemy fights for his pay.
> The enemy fights to defend Colonialism.
> The enemy does not have the full support of a Nation behind him.
> You know why you are fighting.
> You are fighting for Sovereignty and National Honour.
> You are supported by the entire Argentine Nation.
> You fight for a just cause.[23]

Argentine soldiers, although well equipped, lacked motivation and training to fight in foreign wars of conquest. Furthermore, they were often led by invisible desk generals and mediocre officers. Why should soldiers want to fight to the last drop of blood?

In this game theory of sovereignty, it had become increasingly clear that the Beagle Channel scenario was meant to be a real war, as long as it was carried out according to Argentina's terms. But the conditions turned out to be unsatisfactory, for the Chileans showed a firm determination to 'overreact', that is, by December 1978, Chile was ready for all-out war. In the opinion of Chilean Rear-Admiral Francisco Ghisolfo, Chile's military *preparedness* constituted a strong deterrent against Argentine aggression.[24] If Chile had been defenceless, like the Falklands in April 1982, there is little doubt that Argentina would have invaded the Chilean Beagle Channel Islands, and other parts of Chile as well. The 'Malvinas' scenario, despite a three-year pause, was closely linked to the Beagle Channel phoney invasion. These two geopolitical objectives were part of the same Argentine plan of conquests. In January 1982, an Argentine military government spokesman writing for *La Prensa*, Jesus Iglesias Rouco, explained that, by linking the Malvinas and the Beagle, the Argentine junta 'appeared ready to push ahead a diplomatic and strategic plan designed to secure for the country a relevant role in the South Atlantic, in economic, geopolitical and military terms'.[25] By March 1982, it became increasingly clear that the defenceless Malvinas offered an ideal scenario for a big invasion parade.

General Benjamin Rattenbach headed a commission to examine the performance of the Argentine armed forces during the Falklands War, which also studied Chile's role during the conflict. The Rattenbach Report confirmed that the Argentine government had taken the following measures in preparation for a hypothetical war with Chile:

> All military plans contemplated a possible British reaction with Chilean intervention or not. The Commander in Chief of the South Atlantic Theatre of Operations had contingency plans and the Strategic Air Command ordered the Southern Air Force to foresee an eventual Chilean response. The Joint High Command expressed the belief that the biggest British threat would be to try to recover the islands by means of a Naval Task Force simultaneously coordinated with a Chilean attack in the TOS (Theatre of Operations South) and ANEA (Strategic Naval Area South).

Once the islands had been occupied, the Rattenbach Report emphasised that 'a Chilean psychological action and radio campaign equating the Malvinas invasion with all Argentinian territorial claims would take place'. The Report stated further that

> On 14 April 1982, the COMIL (Military committee) analysed the actions and means that the Operational Commands had to take during Preparatory Phase-1 of DEMIL (Military Strategy Directive) No. 1/81 (Chile Case). The commanders-in-chief had decided that each force had to develop its appropriate readiness. Chile deployed and reinforced its forces in the southern region, which coincided with the Argentine

TOS. The Chilean Ambassador in Buenos Aires assured Argentina that 'she had her back covered.' A British helicopter was destroyed by its crew near Punta Arenas. Later, the crew was found and evacuated to Britain without problem, and the crew was awarded medals.[26]

The purpose of this Sea King sortie was to eavesdrop on Argentina's air attacks against British ships. Since the British lacked an airborne early warning system, on 17 April, HMS *Invincible* launched a Sea King helicopter of squadron 846 carrying three crew members and eight Special Air Service (SAS) men equipped with electronic gear to provide covert early warning on Argentina's air raids. The SAS men were landed in Argentina and provided vital information to the Royal Navy ships until the surrender of the Argentine forces, when they were recovered by a submarine. The Sea King could not return to the *Invincible* because of bad weather, and as a result, it had to land in Chile, where it was destroyed by the crew. The official story disseminated by the media was that the helicopter had lost its way and was forced to crash-land in Agua Fresca, near Punta Arenas.[27]

The Rattenbach Report also claimed that the Chilean armed forces' activity in the south hampered Argentinian deployment of part of its forces in the South Atlantic theatre of operations and in the southern theatre of operations. The deployment against Chile had priority over the assignment of army reinforcements to the Malvinas. For this reason, marine infantry brigades VI and VIII were not sent there. Some high-level commands had assessed that Chile could intervene in the conflict. This was apparently considered beyond the systematic analysis of available information. However, Chile collaborated in the salvage of the *General Belgrano*'s crew. The Rattenbach Report concluded that 'the Chilean case deserved special consideration in any circumstance that would require the use of Argentine military power because Chile's actions caused uncertainty regarding its real intentions. Initial predictions prepared at the level of strategic military command were correct and tuned to reality.' The Commission pointed out that the strategy should have been changed when the largescale British reaction became clear, and it was decided to confront it, but this did not happen. Furthermore, the Report claimed that

Argentina's leadership refused to relinquish the two-front war hypothesis, although it faced all the power of the United Kingdom, and had limited resources at its disposal. The Chile threat forced Argentina to divert forces away from the Malvinas front, which caused considerable difficulties to Argentinian combat strength. Finally, the Report asserted that the military junta should have postponed the confrontation with Britain because of Argentina's unfitness to wage war simultaneously on two fronts. Consequently, the initial plan (the invasion of Chile) should have been implemented, or a diplomatic solution to the conflict with Chile should have been found first.[28]

In Argentina, the Chilean Falklands factor played a key role during the war, and continued to have an important part in the trial of the members of the military junta. General Galtieri's defence lawyer, Mr Alberto de Vita, clearly identified the Chile factor during the trial, and also showed that it was a useful scapegoat to justify an untenable military strategy.[29] The Argentinian military were still so obsessed with the Chile factor that they turned Chile into a hypothetical enemy out of all proportion with reality. On 14 April 1982 Admiral Anaya, in a report based upon Argentine naval intelligence, claimed that the Chilean fleet had orders to be ready to support Britain in the Falklands War as of 19 April, and that the Chilean army was secretly deployed at the frontier.[30] It is evident that if part one of the two-pronged *Operativo Rosario* had been successful in its attempted annexation of the Falklands, part two, the invasion of neutral Chile, would have followed. The Brazilian journalist Alexandre Garcia of *Manchete* has also confirmed, based upon first-hand intelligence information, that Argentina had a plan to invade the Beagle Channel Islands, following the Falklands invasion, as part two of *Operativo Rosario*. Under this plan, the strategy would have been, as in the Falklands, to occupy in order to 'negotiate' later.[31] General Galtieri had threatened, in a speech addressed to the Plaza de Mayo masses, that the occupation of the Falklands was only the 'first step in the recovery of the "Argentinian" Atlantic islands'.[32]

The Chilean government hastily adopted a policy of neutrality towards the conflict the very day of the invasion of the Falklands, 2 April 1982.[33] In retrospect, Chile could have chosen one of three possible policies. First, it could have supported the United Kingdom, a traditional friend with strong interests in Chile since the early nineteenth century. Secondly, it could have considered supporting Argentina, a hostile and volatile neighbour, which played an ominous destabilising role in the Southern Cone of South America between 1978 and 1982. The Chilean Ministry of Foreign Affairs opted for a third alternative, which was neutrality. Chile's declaration of neutrality in the conflict was a mistake because it unwittingly told Argentina, although it was not seriously believed: 'You have your back covered'. The architects of such a policy believed, incorrectly, that by remaining neutral in the conflict, Chile would not be making a concession to Argentina. Indeed, it might have been more expedient for Chile to adopt the non-belligerent position which the United States was eventually forced to assume. Chile could have offered the British the use of Chilean lines of communication and supply facilities for their forces, as complements to Ascension Island and Freetown, Sierra Leone, in exchange for modern military equipment. The expected results of such a policy would have neither improved nor worsened relations between Chile and Argentina, since a state of neither peace nor war had persisted for a century.

In return for Chile's neutrality during the Falklands War and its continued support for Argentina's flimsy historical and legal claims against the Falklands, Argentina has systematically backed a Bolivian claim for an outlet to the Pacific Ocean through Chilean sovereign territory, whenever the subject is discussed or voted upon in international organisations. This Argentine position has obvious geopolitical objectives. Chile, on the other hand, has preferred to ignore the principle of reciprocity in its labyrinthine relations with Argentina.

As well as Chile's declared policy during the conflict, there is the question of military co-operation to consider. Considerable exchanges of military information took place between Chile and the United Kingdom, including Chilean purchases of British military equipment, such as warships, airplanes and surface-to-air missiles. After 1979, military relations between Chile and Britain were strengthened as perhaps never before in Chilean history. The well-informed British journalist Duncan Campbell has contended that during the Falklands War there was an Anglo-Chilean understanding whereby Britain gained access to the Punta Arenas air base for RAF spy planes, disguised in Chilean markings. This understanding gave Britain the use of

Punta Arenas and of other areas to infiltrate SAS special forces for secret missions into Argentina. Campbell also claimed that a complete exchange of intelligence, including monitoring and codebreaking of Argentine signals, was carried out by Chilean naval intelligence staff. In exchange, the Chilean government gained Canberra aircraft used in secret operations, which were turned over after the war ended, and a squadron of Hawker Hunters. Finally, Chile was offered a Magnox nuclear reactor, plus supplies of enriched uranium. Restrictions on arms sales were dropped and support was provided at the United Nations Human Rights Commission.[34]

Campbell has also pointed out that at the Punta Arenas airport all the windows of the civil air terminal were whitewashed, with guards posted to prevent anyone looking through. Indeed, at the time Chile was facing a potential aggressor, and a number of strict security measures had been taken at the civil air terminal. However, if at any time there were British planes, it would have been highly unlikely that they would have been stationed in the civil air terminal area, even if they had carried Chilean markings. The truth is that after years of neglect, Chile was engaged in an accelerated programme of building up and reinforcing its air bases and defences. Punta Arenas, being one of the principal target areas of the Argentine Air Force, was being reinforced by important air defence work that justified these severe security measures. If Chile provided assistance to the British during the aggression against the Falkland Islands, it certainly owed apologies to no one.

It may be that 'more nonsense has been spoken about the Islands by those who have never visited them than probably about any other country'.[35] This is why, after having spent time in the Falkland Islands, I am always puzzled by the thriving production of literature about a fictional place called the 'Malvinas'. The reality Falkland Islanders have been living for generations goes far beyond the writing of Malvinas fiction, whether or not it is tinted with historical, legal or geological varnish. The islanders' right to live in their own country seems to be of no concern to those who defend Argentina's geopolitical ambitions. There seems to be a certain brand of political despotism which believes that the right of self-determination is only reserved for those associated with the non-aligned movement. Nevertheless, support for Argentina in the non-aligned movement during the Falklands conflict was anything but enthusiastic, as expressed by A.W. Singham and S. Hune:

> Argentina's presence and record of activity had been singularly undistinguished. Nor was Argentina a popular member of the Movement. The aloofness, arrogance, and racism of the Argentines was well known. Their frequent contention that Argentina was a superior nation because of the purity of their European 'stock' devoid of Amerindian and African 'blood' chafed many a Caribbean and Latin American representative, leading some to hope silently for the deflation of the Argentine ego at the hands of the British.[36]

The much-exploited Argentine anachronistic theme of British colonialism in the Falklands can best be understood with the assistance of General Galtieri. The Italian journalist Oriana Fallaci interviewed Galtieri the day before the Argentinian dèbácle in the Falklands:

> But why do you say colonisation? Isn't it a fact that the 1 800 persons who lived in the Falklands – pardon, the Malvinas – were and are British citizens? Whom did the British colonise in this case? The penguins?
>
> But speaking of colonialism, tell me. You are Italian, aren't you?:
>
> Yes, I am of Italian descent. My grandparents were Italians. My grandmother [was] from Genoa and my grandfather from Calabria. They came here as poor workers with the wave of immigration that took place at the beginning of this century.
>
> So you are colonisers, too. What I mean, Mr. President, is that this country did not belong to the Spanish, to the Italians, to the Germans, the English, the French, the Europeans, the whites who came here and wiped out the natives, massacring them until the last tribe, as was not done even in North America. So why do you call the Falklands – pardon, the Malvinas – colonies?[37]

General Galtieri failed to explain why he – the grandson of Italian immigrants – or his compatriots should have better rights to the Falklands than the descendants of British settlers who have lived there for over 150 years.

The Chilean historian Oscar Espinosa-Moraga claims that one of Chile's peculiar traits is to distrust its own real potential as a nation. For more than a century, Chile has obstinately sought Argentina's friendship, in the false belief that if Argentina were granted an outlet to the Pacific Ocean through Chilean ports, war would be averted.[38] Chile has had a long history of making concessions to Argentina or, according to the same writer, of making territorial quota payments in exchange for peace. Chile is a *status quo* country and has had a long tradition of respect for international law and a rigid, and somewhat naive, belief in the sanctity of treaties. Nothing would have been more remote from its intentions than to attempt a military adventure beyond its boundaries either in 1978 or in 1982. Even when such opportunities were presented to the Chilean armed forces, Chile did not fall into the war trap. What the Argentinian military junta never understood was that Chile had neither claims nor ambitions over Argentine territories, and that the so-called expansionist Chile was only a product of their imaginations.

The lessons of the Beagle Channel quasi-war and of the Falklands War are unmistakably clear for Chile. It needs to continue investing the necessary resources in a strong military machine, as it has been doing since 1978, for that is the most persuasive deterrent against Argentine expansionism. A policy of non-belligerency would have been a more balanced approach instead of neutrality, and a more appropriate policy to adopt towards an openly hostile neighbour in an otherwise difficult situation. It is in Chile's national interest to have a closer *entente* with Britain on Falklands and Antarctic affairs. The history of recent conflicts in the region teaches that well-intentioned policies of peace at any price can result in unexpected and unwanted wars.

NOTES

1. Duncan Campbell, 'The Chile Connection', *New Statesman* (London) 25 January 1985.
2. Colonel Romulo Félix Menéndez, *Las Conquistas Territoriales Argentinas* (Buenos Aires: Circulo Militar, 1982) p. 11.

3. Ibid., p. 292.
4. General Osiris G. Villeagas, *El Conflicto con Chile en la Región Austral* (Buenos Aires: Editorial Pleamar, 1978, Anexo 1, Argentina [La geografia en retroceso]).
5. Jaime Eyzaguirre, *La Soberania de Chile en las Tierras Australes* (Santiago: Zig-Zag, 1958), Article 3 of the Treaty, p. 47.
6. Jaime Eyzaguirre, *La Frontera Histórica Chileno-Argentina* (Santiago: Editorial Nascimento, 1962) p. 24.
7. Oscar Espinosa-Moraga, *El Destino de Chile (1541–1984)* (Santiago: Esparza, 1984) p. 119.
8. Francisco A. Encina, *Historia de Chile* (Santiago: Vea, 1989) Vol. 50, p. 62.
9. Espinosa-Moraga, *El Destino de Chile*, p. 168.
10. J.K. Bluntschli, *Theory of the State* (Oxford: Clarendon Press, 1885) p. 229.
11. Michael Akehurst, *A Modern Introduction to International Law* (London: Unwin Hyman, 1987) p. 15.
12. 'Argentina Declaró Nulo el Laudo', *El Mercurio* (Santiago) 26 January 1978.
13. Virginia Gamba-Stonehouse, *Strategy in the Southern Oceans* (New York: St. Martin's Press, 1989) p. 101.
14. 'Ambiente Bélico en Puerto de Ushuaia', *El Mercurio* (Santiago) 14 December 1978.
15. Colonel Sir Thomas H. Holdich, *The Countries of the King's Award* (London: Hurst and Blackett, Limited, 1904) p. 176.
16. 'Historia Secreta de la Guerra que Evitó el Papa', *Somos* (Buenos Aires) 4 March 1987.
17. 'Prelate of Peace', *The Economist* (London) 13 January 1979.
18. Andrés Oppenheimer, Associated Press cable, United Nations (New York) 21 December 1978.
19. Chilean Admiral, personal interview, December 1989.
20. 'Revela Pinochet que Podria Poner 500 Mil Hombres bajo las Armas', *Diario Las Américas* (Miami) 26 August 1977.
21. Associated Press cable, (Santiago) 27 December 1978.
22. O.R. Cardoso, R. Kirschbaum and E. van der Kooy, *Malvinas: La Trama Secreta*, (Buenos Aires: Sundamericana/Planeta, 1983) pp. 364–5.
23. John Smith, *74 Days: An Islander's Diary of the Falklands Occupation* (London: Century Publishing, 1984) p. 251.
24. Rear-Admiral (R) Francisco Ghisolfo Araya, *Origen y Desarrollo del Diferendo Limitrofe Austral* (Santiago: Instituto de Investigaciones del Patrimonio Territorial de Chile, Universidad de Santiago, 1983) p. 35.
25. 'Southern Notebook', *Latin America Regional Reports Southern Cone* (London) 9 April 1982.
26. *Informe Rattenbach* (Buenos Aires: Ediciones Espartaco, 1988) p. 191.
27. David Brown, *The Royal Navy and the Falklands War* (Annapolis: Naval Institute Press, 1987) p. 169.
28. *Informe Rattenbach*, p.192.
29. Raúl O. Garcés, Associated Press cable (Buenos Aires) 28 July 1988.
30. Ibid.

31. 'Argentina Pensaba Repetir Invasión en Canal Beagle', *La Prensa Austral* (Punta Arenas) 13 May 1982.
32. Agustin Oyarzún Lemonnier, *Guerra en las Falklands* (Santiago: Editorial Cumbres, 1983) p. 378.
33. Carlos M. Goni Garrido, *Crónica del Conflicto Chileno Argentino* (Santiago: Ediar Editores, 1984) p. 176.
34. Duncan Campbell, 'The Chile Connection'.
35. Ewen Southby-Tailyour, *Falkland Islands Shores* (London: Conway Maritime Press 1985) p. 1.
36. A.W. Singham and Shirley Hune, *Non-Alignment in an Age of Alignments* (London: Zed Books Ltd., 1986) p. 241.
37. 'The Argentine General Who Never Fought in a War', *The Washington Post*, 13 June 1982.
38. Espinosa-Moraga, *El Destino de Chile*, p. 162.

4 The Role of the Falkland Lobby, 1968–1990

Clive Ellerby

In the mid-1960s pragmatism would have probably led the British government to transfer the sovereignty of the Falklands had not the Falkland Lobby arisen to counter the pressure of Argentine demands. It is, therefore, surprising that the activities and influence of this pressure group have been largely overlooked or misunderstood. The Franks Report gave only one specific reference to the Falkland Lobby and did not elaborate on how it had affected the British government's perception of the dispute.[1] James Callaghan, the former Labour Prime Minister, asserted erroneously in his memoirs that the Falkland Lobby was concerned only with the interests of the Falkland Islands Company.[2] One recent work on the conflict has a useful chapter on the Lobby, but it is both brief and based principally on official records.[3] In most accounts the Lobby is only a shadowy, perhaps demonic, presence.

The Falkland Lobby was a broad alliance of individuals, commercial interests, sympathetic MPs and Falklanders who were united by the objective of upholding the islanders' right to self-determination. What was its importance in the development of the conflict after 1968? The closure of official records under the thirty years rule presents a major obstacle to research. However, it is not insurmountable. The hitherto untapped archives of the United Kingdom Falkland Islands Committee (UKFIC) and the Falkland Islands Association (FIA), in particular, provide a fascinating insight into how a successful pressure group perceived the dispute and influenced events. These archives may be supplemented with private papers and the records of the Falkland Islands Company (FIC). The sources are richer than might be supposed.

The perceived value of the Falklands to Britain is a vital issue in understanding both British government policy towards the islands and the aims of the Falkland Lobby. The islands were annexed in 1833 to give Britain a secure port on a major trade route around Cape Horn and a base for the Royal Navy in time of war. The preoccupation with the development of Port Stanley shaped the approach to the colonization of the islands. The land was divided into vast ranches which were leased and sold to the Falkland Islands Company, which was granted a Royal Charter in 1851 to initiate development, and to a handful of settlers. In the short term the Falklands prospered and were granted Crown Colony status in 1892. However, in the long term underdevelopment was perpetuated until the 1980s. The form of landownership and reliance on a monocultural economy inhibited further development. A more intensive use of the land and the creation of new industries required an improved infrastructure. This was not considered necessary since profitable sheep farming needed the minimum of investment. The population remained static at about 2,000 because sheep farming could not support greater numbers. The British government perceived the colony as a strategic outpost with only extrinsic value.[4]

By the mid-1960s the Falklands appeared to have become a colonial anachronism and a liability as a result of a combination of factors. These included the withdrawal from empire, pressure from the United Nations and the reappraisal of British defence policy. The Falkland Lobby was formed in response to concern that the British government planned to deny the islanders' right to self-determination by transferring the sovereignty of the Falklands to Argentina. British and Argentine diplomats initiated confidential talks in July 1966. In March 1968 Michael Stewart, the Foreign Secretary, told Parliament that the government wanted to resolve the sovereignty dispute in the interests of the islanders, but he was ambiguous about the importance of their wishes.[5] A 'Memorandum of Understanding' was agreed in August 1968. It said that the Falklands would be transferred once an agreement was reached on how the promised Argentine guarantees would be able to secure the islanders' interests. The publication of the Memorandum was to be accompanied by a unilateral British statement that the handover should only take place if it were acceptable to the islanders.[6]

The development of representative government in the colony played an important role in legitimizing the Lobby's campaign. Elected representation on the Legislative Council was first introduced in 1949. A new constitution in 1964 gave both the

Executive and Legislative Councils clear unofficial majorities. As the sovereignty talks with Argentina progressed, the Colony's Executive Council was sworn to an oath of secrecy and shown early drafts of the Memorandum.[7] The islanders were apprehensive that the secret talks would result in a *fait accompli*. On 27 February 1968 the unofficial members of the Executive Council – Arthur G. Barton, Dick Goss, Sydney Miller and G. C. R. Bonner – wrote to every MP and to the national press. The letter began:

> ARE YOU AWARE THAT –
> Negotiations are now proceeding between the British and Argentine Governments which may result at any moment in the handing-over of the Falkland Islands to The Argentine.
> TAKE NOTE THAT –
> The Inhabitants of the Islands have never yet been consulted regarding their future – they do NOT want to become Argentines – they are as British as you are, mostly of English and Scottish ancestry, even to the 6th generation – five out of six were born in the Islands – many elderly people have never been elsewhere – there is no racial problem – no unemployment – no poverty, AND WE ARE NOT IN DEBT.[8]

The manifesto concluded with an appeal for help.

The Falkland Lobby was launched by Arthur Barton who visited Britain to publicize the islanders' cause. Barton was a former colonial manager and director of the FIC who had settled in the Falklands in 1920 and married into one of the oldest pioneering familes. Frank Mitchell, the Company Secretary of the FIC in London, arranged interviews with the media and parliamentarians.[9] A parliamentary debate began on the government's policy towards the Falklands on 26 March 1968. The night before, Barton had addressed a meeting of the United Kingdom Branch of the Commonwealth Parliamentary Association. The most prominent speakers in the debate were the Conservative back-benchers John Biggs-Davison, Michael Clark Hutchison and Bernard Braine, and the Labour MP Clifford Kenyon.[10] The interest of these MPs was the result of their various concerns. For example, Clifford Kenyon had visited the Falklands in March 1967 as a member of a Commonwealth Parliamentary Association delegation, while John Biggs-

Davison had protested vigorously against the abandonment of Empire and was devoted to the cause of Ulster Unionism.[11] It seems that the emergent lobby was motivated by a combination of imperial nostalgia, a sense of obligation to people who wished to remain British, and resentment against the United Nations for appearing to ignore the wishes of peoples in the interests of decolonization. Clive Christie has concluded succinctly that '"Honour" rather than national interest was at the root of the Falkland Lobby's position'.[12]

The Falkland Lobby was formalized on 25 March 1968 by the formation in London of the so-called Falkland Islands Emergency Committee. It was instigated by Bill Hunter Christie, a barrister at Lincoln's Inn. Christie was Third Secretary at the British Embassy in Buenos Aires between 1946 and 1948. He developed an interest in the Falklands and subsequently wrote an authoritative history of Antarctica.[13] He said in 1987 that his dedication to the Falkland cause could be explained by a reference made by Lord Shackleton a decade earlier. It was thought that the Falklands could be the moral making of Britain in both a physical sense (through economic development which would revive a pioneer instinct) and the reversal of Britain's decline by upholding certain principles. This view, based on a belief in the unique contribution of the British nation to the progress of humanity as a whole, has similarities to the ideas of the famous British imperialist Cecil Rhodes. Christie approached Patrick Ainslie, the Chairman of the FIC, and suggested calling a meeting of interested individuals.[14]

A meeting was held at the FIC's London office to consider the manifesto sent out by the Executive Councillors. It was attended by the directors of the FIC, Barton, Mitchell, Christie, Falkland landowners and farm managers, the Labour MP Clifford Kenyon and the Conservative MP John Smith. It was decided to form a committee – Mitchell was appointed Secretary and Sir John Barlow, Bt. (a Director of the FIC and former MP), Chairman. Christie persuaded the meeting that the committee should not be partisan (despite overwhelming Tory support) and should have Labour and Liberal representation.[15] The Falkland Islands Emergency Committee's main activity was raising public awareness. Sympathizers in Parliament united to form the 'United Kingdom/Falkland Islands All Party Group of the Commonwealth Parliamentary Association'.[16]

The first achievement of the Emergency Committee was successfully to counterbalance Argentine pressure on the British government. The campaign by the Emergency Committee culminated in the Cabinet's decision to abandon the Memorandum of Understanding on 11 December 1968. There were two important factors which contributed to this success. First, the press was rife with rumours. The newspapers became increasingly critical of government policy after Lord Chalfont, a Foreign Office minister, returned from a mission to the Falklands in November. For example, the *Daily Telegraph* announced 'BRITAIN READY TO SURRENDER FALKLANDS', and claimed that Chalfont had left the islanders in a 'depressed and nervous state'.[17] Secondly, criticism in Parliament of the government's policy mounted because the Conservative Shadow Cabinet adopted the cause of the Falkland Lobby. S.E. Finer has argued that the policies of political parties are often 'framed in opposition' when they 'have no civil servants to advise them'.[18] On 28 November the backbench '1922 Committee' of the Conservative Party sponsored a motion which was signed by more than one hundred Conservative MPs. It called on the Prime Minister 'once and for all to make clear that the Falkland Islanders are British, will remain British, and need not fear any transfer against their will to an alien land'.[19] The Emergency Committee supplied Michael Clark Hutchison and others with material for Parliamentary questions and prepared a circular on non-party lines.[20] On 9 December Christie wrote to Jim Prior, the Parliamentary Private Secretary to the Leader of the Opposition, with notes on the Falklands, in case 'this is a front Bench matter'.[21] On 12 December Sir Alec Douglas-Home, the Foreign Affairs Spokesman, pledged that a Conservative Government inheriting negotiations with Argentina would 'strike the question of sovereignty from the agenda'.[22] The Emergency Committee appeared to have both generated opposition to government policy on the Falklands and gained a commitment from the Conservative Party.

For Edward Heath, the Falklands issue was probably a minor skirmish in a war to discredit the Labour Party and win the next general election. He was more interested in the vision of a 'greater Britain in a greater Europe' than residual imperial commitments or a sense of 'kith and kin'. In March 1966 Labour achieved a crushing victory over the Conservatives, winning 363 seats to 253. Heath was compared to Hugh Gaitskell as a 'born loser'. The Conservative Party was a broad coalition which faced the problem of fragmentation. The Rhodesian crisis split the Tories between the pro-Ian Smith Monday Club on the right and the anti-Smith, anti-racists on the left. When the government decided on oil sanctions, Heath and the Shadow Cabinet urged abstention. Nevertheless, Julian Amery and fifty MPs voted against oil sanctions while thirty-one Tories voted in favour. In April 1968 Enoch Powell threatened to tear the Party apart with his notorious Birmingham speech on the consequences of uncontrolled Asian immigration.[23] His use of rhetoric and emotion were very extreme: for example, he described a single old white lady, who was abused by her immigrant neighbours, and had 'excreta pushed through her letterbox'.[24] The Falklands issue presented Heath with an opportunity to unite the Party, divide Labour and recapture public belief in the Conservative Party's ability to defend British values. Heath stated before the Royal Commonwealth Society that 'the Government gives the impression it is sorry the islanders wish to remain British. But this is a matter for pride, not for shame'.[25] This attack added to the pressure from Labour backbench MPs, such as Clifford Kenyon, who opposed the government's line on the Falklands.

The government created a dilemma for itself by initiating sovereignty negotiations with Argentina while promising the islanders that the Falklands would not be transferred. British policy was given a degree of continuity under both Labour and Conservative governments by a predominant view in the Foreign Office that a transfer of sovereignty was inevitable. It was assumed that the Falklanders could be assimilated by the essentially European mainland population. A new initiative was launched in 1970 to reduce the islanders' dependence on Britain for essential services and supplies. The intention was to persuade the Islanders that a transfer of sovereignty was at once in their interests and inevitable. David Scott, the Under-Secretary in charge of dependent territories at the Foreign Office, said that his principle was, 'rape of the Falklands, no; seduction by all means'.[26] The 1971 Anglo-Argentine Communications Agreement sought to improve relations with the provision of a weekly air service to Patagonia and scholarships for islander children to attend schools in Argentina. In 1974 the Argentine State Oil

Company was granted a monopoly to supply the Falklands with certain petroleum products at mainland prices.[27]

The islanders did not resist this gradual approach because of the colony's continued economic difficulties. In December 1970 the FIC announced that it intended to withdraw the *RMS Darwin,* which had provided the colony's principal external link.[28] The ship carried mail, passengers, fuel supplies and fresh food on monthly round trips to Montevideo in Uruguay. The Falkland Islands Emergency Committee stood down in 1971 following an assurance of satisfaction with the Communications Agreement from the Colony's Legislative Council. Bill Hunter Christie commented: 'We thought that our job was done'.[29] Arthur Barton privately believed that integration of the Falklands into Argentina was inevitable.[30]

The Falkland Lobby was relaunched in response to the end of the Anglo-Argentine rapprochement which followed Argentine internal disorders and the return of a nationalistic Peronist Government in 1973. The Falkland talks had raised Argentine nationalist expectations which were mirrored in demands for the immediate transfer of sovereignty. In July 1972 the Emergency Committee was reorganised. The members agreed that the Committee should be enlarged and placed on a 'permanent footing'. [31] Christie and Mitchell became Joint Honorary Secretaries. It is important to note that the Committee's links with Parliament were strengthened. Letters were written to each of the larger political parties inviting them to appoint a member to serve on the Committee. MPs from the Conservative, Labour, Liberal and Scottish Nationalist Parties subsequently joined. Among the most influential MPs who later became members were Sir Nigel Fisher, the former Conservative Parliamentary Under-Secretary of State for the Colonies, Ted Rowlands, a Labour Foreign and Commonwealth Office Minister during the 1970s, and Russell Johnson, the Liberal Party Spokesman on Foreign Affairs. In April 1973 the Committee adopted the following objective: 'To assist the people of the Falkland Islands to decide their own future for themselves without being subjected to pressure direct or indirect from any quarter'.[32] In July 1974 the Committee held a reception at Lincoln's Inn. The presence of Earl Mountbatten of Burma as special guest reflected the British government's approval of the campaign.[33]

However, the self-appointed Emergency Committee lacked

authority in representing the islanders. In May 1974 it agreed in principle that a sister Committee should be formed in the Colony.[34] Christie suggested that it should act as 'a two-way link with us in London'.[35] In October a ten-member Falkland Islands Committee (Local Branch) was formed at a public meeting in Stanley.[36] Christie also proposed that the renamed United Kingdom Falkland Islands Committee (UKFIC) should fulfil 'some of the functions of a Falkland Islands High Commission in London – something which the Colony cannot afford or justify if only because of its size'.[37] The UKFIC wanted to retain its independence as a group of supporters rather than to become subservient to the Islanders. This created an anomaly. The UKFIC sought to perform the functions of an embassy but was not accountable to either the Legislative Council or to the sister committee. It believed that government policy towards the Falklands was formulated as well as implemented by the Foreign Office. It was suspicious that officials based their actions, 'on the premise that the existing political situation is temporary whereas decisions should be made on the basis of a permanent link with Britain'.[38] Christie thought that the UKFIC and the Foreign Office should collaborate since they shared an identical objective. Nevertheless, in 1975 Robin Edmonds, the Superintending Under-Secretary of the Latin American department, rejected a proposal that the UKFIC act as an advisory body to the Foreign Office.[39] A distrust developed.

The 1973 oil crisis was a turning-point in that it focused attention on the colony's resources. The fears about fuel shortages in the early 1970s enabled the Falkland Lobby to see the potential of the colony and made it harder for the Foreign Office to dismiss the importance of the Falklands. By 1975 the Falkland government had received at least forty-nine speculative applications from oil companies of various nationalities for onshore and offshore prospecting and drilling licences. Several companies carried out survey flights while two parties of geologists, disguised as bird watchers, worked ashore.[40] In 1975 the US Government Geological Survey Bulletin reported that the Argentine continental shelf held an estimated 200 billion barrels of oil.[41] The Foreign Office opposed any exploration around the colony until its own studies were completed.

The second major achievement of the Falkland Lobby, after it had won an assurance from the British government to respect the islanders' wishes, was to initiate a debate about how economic development could transform the colony into a valuable asset for Britain. In November 1974 David Ainslie, a London solicitor with close family ties to the FIC, argued that intensive economic development was required to clarify and strengthen the unique British identity of the community in the islands. The bulk of the islanders' energy and food requirements could be provided by the development of oil resources, agriculture and fishing. A self-sufficient population of 10,000, which was able to justify and supply vital services, could win international support for the right to self-determination.[42]

Much of the drive behind the pressure group came from the enthusiasm and commitment of Bill Hunter Christie. After a visit to the Falklands in January 1975 Christie also expressed his belief that radical change was necessary. Christie argued that the Falklands had a huge potential for development and drew parallels with other island communities such as the Faroes, the Shetlands and the Canaries. In 1988 he claimed that the potential of the Falklands was similar to Tasmania which supported nearly 450,000 people.[43] It is extremely difficult to comment on the colony's full potential, but such speculation undoubtedly helped to enliven the debate. Christie recommended a comprehensive economic review of the colony and the establishment of a development agency. Aspects of his vision included the creation of owner-occupied farms, a road network, an international airport, a fishing industry and afforestation. Christie stressed that, 'With or without oil, the Islands are an asset of great potential value to Britain and indeed it may be to Europe as well'.[44]

By the mid-1970s the Falklands dispute had reached an impasse due to inflexibility on both sides. Attention in the dispute then shifted from the issue of sovereignty and improving relations between the islands and the mainland to the economic resources of the south-west Atlantic. The 1976 Shackleton Report symbolised this turning-point. It was the first comprehensive socio-economic survey of the islands which considered how to direct development aid as well as the Falkland Government/FIC relationship, and Falkland politics. The survey appears to have originated from two simultaneous proposals to the British government in May 1975. The first was made by the colony's Executive Council when it met David Ennals, a Foreign and

Commonwealth Office Minister, in Rio de Janeiro.[45] Secondly, the UKFIC was also instrumental in the decision to conduct a survey. On 8 May 1975 it attended an informal seminar with Ennals at the Foreign Office to enable an exchange of views on diverse subjects including oil exploration, fisheries and the constitution.[46] In October 1975 the Labour peer Lord Shackleton, Deputy Chairman of the Rio Tinto Zinc Corporation and son of the renowned Antarctic explorer, was appointed Chairman of a survey team and the Economist Intelligence Unit was commissioned to recruit the relevant experts. The decision to appoint an independent study was probably intended to counter any criticism that its findings would be influenced by the government. It is difficult to assess the expectations of the interested groups with regard to the survey. The Foreign Office perhaps wanted the report to support the argument that integration with Argentina was inevitable, while the Falkland Lobby and the islanders hoped it would show the potential for development independent of Argentina.

The Shackleton Report, as it was called, was presented to the Foreign Secretary in May 1976. At a news conference Shackleton rejected the idea that development was doomed without Argentina collaboration but thought it would be 'foolish' not to seek it.[47] The report presented three important innovations. First, it advocated Argentine involvement in exploiting the offshore oil, fish and krill resources of the Falklands and Dependencies. However, the recommendations that dealt with internal development sought to strengthen the colony's economy in 'splendid isolation' from Argentina. The key recommendation was the extension of Stanley Airport's runway (which was built by the British government as part of the 1971 Communications Agreement). The team recognised the need to bypass Argentina's stranglehold on external travel, establishing regular air services with Montevideo and Punta Arenas in Chile. The report's second innovation was to conclude that local initiatives towards development and Falkland politics were inhibited more by dependence on absentee landowners for employment, housing and provisions than by either isolation or the sovereignty dispute. A number of recommendations were made for mainly social reasons, such as the wider ownership of land. The third unexpected conclusion was that private investments and property income had historically flowed from the Falklands to Britain. This amounted to over £5 million (in current prices) over the past twenty years.[48]

On 2 February 1977 Anthony Crosland, the British Foreign Secretary, announced that the government acknowledged the potential for development in the Falklands but would not fund the major capital projects recommended by the Shackleton Report because of the recession.[49] It seems that the government's stress on co-operation with Argentina as a precondition for development reflected surprise at the outcome of the report. Ted Rowlands, a Foreign and Commonwealth Office minister at the time, recollected that the recommendations were unexpected. Shackleton was thought to have 'lost his heart' to the islanders.[50] The total cost to the British government for implementing all the recommendations was £12.9 million over five years, of which £7.5 million (towards the development of the fisheries) could be deferred. Therefore, the British government's contribution to the funding of all the other recommendations was £5.4 million. The most expensive project was the airport runway extension (£3.5 million).[51] The Shackleton Report was based on the assumption that the Falklands would remain a British colony. It presented a strategy to maintain and strengthen the position of the islanders which, from a Foreign Office perspective, appeared both expensive and highly provocative to Argentina.

After the publication of the Shackleton Report in July 1976, the Falkland Lobby decided that a public relations campaign on behalf of the islanders was imperative. Christie said that, 'If the Government does not resolve in principle to implement the broad lines of Lord Shackleton's recommendations and does not decide to extend the airfield, I consider that sovereignty will pass to Argentina within five years'.[52] In March 1977 the 'Falkland Islands Research and Development Association' (FIRADA) was incorporated as a non-profit making business. Its purpose was both to guide economic development and to act as an umbrella organisation to represent the islanders' interests in Britain.[53]

In January 1977 the 'Falkland Islands Office' was opened in Westminster. Sallingbury Ltd., Parliamentary and public relations consultants, were retained to mount a lobbying exercise in support of the UKFIC to ensure that the Shackleton Report was not ignored. The campaign involved the development of existing Parliamentary support into an ongoing lobby with clearly defined objectives, contact work with government officials and industrialists, and informative articles in the media to raise public awareness.[54] In April 1978 Air Commodore Brian Frow was appointed to the salaried post of Director General of FIRADA. Frow had earlier served as a defence attaché in Moscow and a member of the Joint Intelligence Staff at the Cabinet Office.[55] By 1979 the Office had written over 2,000 letters, distributed some 6,000 pamphlets and leaflets, issued several hundred press releases and maintained close links with the media.[56] In February 1979 FIRADA gave Sallingbury a directive to put more emphasis on the future prospects of the Falklands, rather than on sovereignty, to attract commercial interests.[57] The June 1979 edition of *The Falkland Islands News* described the islands as 'Britain's new North Sea' because the development of deep-sea fisheries, alginates (seaweed products) and oil would be of direct benefit to the British economy.[58] The Falkland Lobby's influence in British politics was based on a large body of support in Parliament. In 1979 an Early Day Motion which called on the government to implement the Shackleton Report was signed by 118 MPs in the House of Commons.[59]

The absence of a rival pressure group which advocated a resolution of the sovereignty dispute gave the Lobby an opportunity to inculcate its ideas about the future of the Falklands in the minds of MPs. But FIRADA faced two major internal problems. The first was insufficient funds. In 1979 the Falkland Islands Office cost about £25,000 while its total income was £22,090 (£2,408 from individual members and donations and £19,682 from corporate subscriptions). The most important subscribers were the FIC, Alginate Industries, the Falkland Sheep Owners' Association and the wool merchants Jacomb, Hoare and Co.[60] The estimates for the year beginning April 1980 showed a £9,000 deficit.[61] FIRADA responded to the shortfall with cutbacks and a fund-raising campaign. The second problem was that the Association lacked credibility in the Falklands. Many islanders were critical of FIRADA's lack of accountability, and the Falkland government refused to contribute towards the costs of the Falkland Islands Office. Adrian Monk, elected member for East Falkland, denounced FIRADA for usurping the authority of the Legislative Council. He stated that, 'we want government here, not government by commercial hacks in London'.[62]

The Falkland Lobby's third significant achievement was to lay the foundations for the colony's post-1982 prosperity based on a South Atlantic fishing industry. The Falkland Lobby instigated the 'South Atlantic Fisheries Committee' (SAFC) with the aim of saving Britain's deep-sea fishing industry and of proving that the Falklands were an asset to Britain. The basic objective was 'to see a fleet of thirty British trawlers based in Port Stanley all the year round, the crews being changed regularly by air and regular freight service bringing the catch back to England'.[63] The loss of traditional fishing grounds as a result of Iceland's extension of its fishing limits caused the demise of Britain's distant-water fishing fleet. In January 1977 the British government established a 200-mile fishing zone around the United Kingdom after it failed to break the Icelanders' resolve during the 'Cod War'. As a result the number of distant-water boats declined from 168 in 1972 to fifty in 1980.[64] On 28 July 1977 a meeting was held at the House of Commons attended by representatives of the fishing industry. It was sponsored by James Johnson,the Labour MP for Kingston-upon-Hull (a constituency seriously affected by the decline in the distant-water fishing fleet) and chaired by Bill Hunter Christie. As a result the SAFC was formed, composed of representatives from various interests such as British United Trawlers, the National Fish Fryers' Federation and the National Farmers Union.[65] The objective of the SAFC was to produce plans for a commercial fishery survey of the seas around all the British south Atlantic territories. In April 1978 it sent a memorandum to John Silkin, the Secretary of State for Agriculture, Fisheries and Food, which proposed that the government fund a twelve-month commercial survey of the south west Atlantic fishery. It claimed that a British fishing industry could depend, 'on fish meal and fish oil for its bread and butter and high value white fish for its jam'.[66] The proposal gained broad support in both Parliament and the press.

It would appear that the SAFC failed to gain any influence over government policy because of the rise of a rival pressure group which represented a different part of the British fishing industry. In June 1980 Alick Buchanan-Smith, the Minister of State for Agriculture, Fisheries and Food, announced that the Government would not fund a survey of the south west Atlantic because future commercial viability was uncertain. He then argued that government aid for the fishing industry 'must be confined to activity in the U.K. and its surroundings waters'.[67]

This reflected the shift in political influence away from the distant-water sector of the British fishing industry to the near-water interests as a result of the widespread introduction of 200-mile national fishing zones. The sharp decline in distant-water vessels was largely counterbalanced by an increase in small and medium-sized vessels which concentrated on grounds within Britain's 200-mile fishing zone.[68]

The loose federation created around FIRADA, with solid Parliamentary support, proved formidable when the issue of sovereignty came to the fore. In July 1980 the Cabinet's Defence and Overseas Policy Committee, chaired by the Prime Minister, reviewed policy towards the Falklands and agreed an attempt should be made to reach a solution of the dispute on the basis of a leaseback arrangement. Nicholas Ridley, the Minister of State at the Foreign Office with responsibility for the Americas, visited the colony in November to assess the level of support for leaseback.[69] He presented the Islanders with three options – a condominium, a freeze of the dispute for about twenty-five years, and a leaseback scheme. The minister stressed the problems of instability with the first two options and the benefits of the third. Leaseback would involve a transfer of titular sovereignty to Argentina in exchange for a lease on the islands and a 200-mile maritime zone which would last at least one generation. If Argentina accepted the suggestion the British administration in the Falklands would be maintained and the external inhibitions on economic development would be removed.[70] The opinion of the islanders appeared to be divided on leaseback, with a substantial minority opposed and the majority undecided.[71]

As is well known, on 2 December 1980 Ridley was savaged in the House of Commons when he reported the outcome of his visit to the islands. The minister's assurances, that any solution had to be endorsed by the islanders and Parliament, were ignored. Peter Shore, the opposition Labour Party spokesman, and the MPs of the Falkland Lobby demanded paramountcy for the islanders' wishes and anticipated their rejection of leaseback. Julian Amery described the proposal as 'profoundly disturbing' while the Liberal MP Russell Johnson criticised the 'shameful schemes for getting rid of these islands which had been festering in the Foreign Office for years'.[72] On 4 December the Cabinet concluded that the issue was highly emotive for Parliamentary and public opinion in Britain. It also thought that the opposition of the islanders had been exaggerated by the 'attitude of their champions in Parliament'.[73]

It is ironic that the supporters of the Falkland cause in Parliament, who had stifled the leaseback initiative, reflected the views of neither the UKFIC nor the islanders. The UKFIC was divided on the merits of leaseback, which were discussed with Nicholas Ridley at a meeting on 3 December 1980. At a meeting on 17 December a number of UKFIC members expressed support for the approach. Ralph Merton, who had represented Alginate Industries Ltd., said that a leaseback agreement was the best way to encourage commercial investment. Frank Mitchell agreed. He warned that 'the Islands might not survive as an economically viable entity for more than possibly 3 years'.[74] A vote on the wisdom of issuing an immediate statement in favour of leaseback was only defeated by six votes to five. The UKFIC then unanimously decided to adopt a neutral stance and press for further information about the attitude of the Argentine government.[75]

The islanders were initially divided on leaseback. In 1983 Patrick Watts, the manager of the colony's broadcasting service, estimated that 50 per cent of islanders favoured leaseback from the many debates held on the radio. He claimed that:

> There were households in Stanley split down the middle . . . Representatives of the Sheepowners' Association told me that they thought it was the answer, representatives of the Employees' Union told me they thought it was the answer. One person who sat before me came back and told me that 97 per cent of the people of the West Falkland would like to talk about it, not that they supported it.[76]

Derek Evans, the elected Legislative Councillor for the West Falklands, claimed that the 'people most in favour of lease-back tended to be managers, probably because they are more acutely aware of the farms' financial positions'.[77] Islander opinion then hardened against leaseback. Watts believed that the turning-point occurred on New Year's Eve when Adrian Monk gave a Churchillian speech over the radio. Watts thought that it 'certainly changed the minds of a lot of people in the Falklands without doubt, and a lot of people will not now admit that they considered leaseback as a possibility'.[78] Monk argued that the representatives of British based companies lacked the Kelpers' stake in the Islands and were primarily interested in short-term profits. He then announced: '*I THINK THE WHOLE CAMPAIGN STINKS*. . . Don't be misled. Don't be worried about the consequences of saying "We are British". Our country will remain British'.[79] On 6 January 1981 the colony's Joint Councils passed a motion which stated that they did not like any of the ideas put forward by Nicholas Ridley but agreed that talks should be held with Argentina. The aim should be to freeze the dispute over sovereignty for a specified period of time – Tim Miller, elected Legislative Councillor for the Camp Divisions, favoured fifty years.[80]

The leaseback initiative revealed that the Falkland Lobby was a fairly loose confederation of three groups which did not always work in unison but which were united by the same objective. The sympathetic MPs in the House of Commons denounced leaseback before either the islanders or the UKFIC had had sufficient time to consider it. A large proportion of both islanders and members of the UKFIC initially thought that the option *should* be pursued with Argentina. However, the UKFIC failed to keep in step with opinion in the Falklands when attitudes hardened against leaseback. In November 1981 Velma Malcolm, the Secretary of the Falkland Islands Committee (Local Branch), advised the UKFIC to reappraise its ambiguous position and to oppose leaseback openly.[81] The Falkland Lobby probably influenced Lord Carrington's decision in September 1981 to reject a more public and active campaign to educate islander and British public opinion.[82] Many in the Foreign Office believed that a precondition for an agreement with Argentina was a campaign by the government to publicise the colony's isolation and the huge cost necessary to deter an Argentine attack. However, Carrington evidently thought that the threat was not urgent enough to justify challenging the Falkland Lobby.

The British government was faced with volatile and unpredictable forces between 1966 and 1982 which inhibited the formulation of a consistent policy to resolve the dispute. The changes in government policy were responses to new situations as they arose. The Falkland Lobby was important because it both countered any attempt to apply pressure on the islanders, and drew attention to the economic potential of the colony.

During the 1982 war the Lobby provided the task force planning team at the Ministry of Defence with intelligence throughout the conflict, using its comprehensive network of islanders and experts in Britain.[83] This assistance was important since detailed information about local conditions was probably scarce. The Office also provided the rudiments of a 'Government in Exile'. Rex Hunt, the Colony's Governor, and Dick Barker, the Chief Secretary, were given a room at the Foreign Office. However, John Cheek, a Legislative Councillor, and Bill Luxton, an Executive Councillor deported by Argentina, used the Falkland Islands office. Islanders around the world offered support and travelled to London. The volunteers supported the government by giving radio, television and press interviews.[84]

After the war, the British government's commitment to the islanders resulted in the implementation of the Falkland Lobby's initial programme for economic development to bolster the position of the islanders. The completion of Mount Pleasant Airport in 1985 gave the islands both a permanent symbol of Britain's commitment and a runway capable of handling the largest aircraft in existence. Regular twice-weekly flights to Britain via Ascension Island were established. The British government also committed £31 million over a six-year period to fund the implementation of a revised Shackleton Report programme. In 1984 a Falkland Islands Development Corporation was created, based on the Scottish Highlands and Islands Development Board, with the broad aim of making the colony a more viable economic entity. The priorities included the diversification of the economy, an improved infrastructure and selective immigration to increase the population.[85] The most important change was the transfer of land in the colony. The proportion of land under local ownership increased from 24 per cent before 1980 to 73 per cent in 1989.[86] The establishment of owner-occupied farms encouraged local initiative towards development and assisted the emergence of a Falkland sense of identity which combined a local patriotism, which went a long way back, with a British culture and reliance on British protection.

The activities of the Falkland Lobby were scaled down as a consequence of the political changes which followed the 1982 Conflict. In January 1983 Sir Rex Hunt, now Civil Commissioner for the Falklands, said that the UKFIC should continue as a political body on behalf of the islanders.[87] In 1984 FIRADA ceased trading and was replaced by the 'Falkland Islands Association', which was formed as an unincorporated supporters' organization. The office in Westminster was retained. The Colony's Legislative Council marked its approval of the changes in the Lobby with a donation of £5,000 for the political work of the UKFIC.[88] The Lobby concentrated on publicity. In July 1982 a glossy brochure was sent to all MPs and interested peers, Members of the European Parliament, foreign ambassadors in London, the United Nations, the entire US Congress, the media and commercial organizations. It claimed that the Falklands had good prospects for development without dependence on Argentina.[89] In the financial year 1986–7 almost one-third of the Association's total expenditure was devoted to the publication of newsletters which stressed the positive aspects of development in the Falklands.[90]

The Falkland Islands government also took over many of the Falkland Lobby's activities. In July 1982 the colony's Joint Councils decided to establish a Falkland Islands Government Office (FIGO) in London. In September the work carried out by FIRADA on development and immigration was transferred to Adrian Monk, who was appointed to the post of Falkland Islands Government Representative in London. His duties included the promotion of the islands in the media and Parliament.[91] Broad Street Associates, a public relations company, was commissioned to provide direct contact with MPs and peers.[92] Representatives of the FIGO regularly attended the annual conferences of both the Conservative and Labour parties.[93]

The period between 1982 and 1989 showed a significant shift in British public opinion towards support for a negotiated compromise on sovereignty with Argentina. The cost of defending the colony was an important factor. By 1988–9 the Ministry of Defence expected that the total annual cost of operating the garrison would stabilize at about £100 million.[94] A Gallup Poll in February 1986, for example, revealed that the public appeared to divide along party lines on the future of the Falklands, with a small majority of Labour and Alliance voters favouring some sort of change.[95] The diminished support for the retention of the islands was reflected in the formation of a pressure group to rival the Falkland Lobby. In 1983 the South Atlantic Council (SAC) was formed to promote a peaceful settlement of the sovereignty dispute. It was composed of politicians from all parties, academics, churchmen and businessmen.[96] Its leading figures were the Conservative MP Cyril Townsend and the Labour MP George Foulkes, who was the Opposition front-bench spokesman on Latin America. The SAC's most notable achievements were to publish papers on various aspects of the dispute, such as options for a possible settlement, and to sponsor the re-establishment of links with Argentina.[97] However, the SAC's influence was limited compared to the Falkland Lobby because it lacked both resources and a clearly defined objective based on a principle.

The Falkland Lobby saw large-scale economic development as crucial if the British public's perception of the colony as a costly liability was to change. A major concern during this period was the inadequate restrictions on fishing around the islands which led to a bonanza for foreign distant-water fishing fleets. The number of trawlers and factory ships in Falkland waters increased from 250 in 1984 to over 600 in 1986.[98] The British government initially favoured a multilateral approach to the management of the south-west Atlantic fishery. However, it came under mounting pressure to revise this policy. First, concern that over-fishing would damage the eco-system of the seas around the Falklands resulted in the formation of a loose coalition to demand the imposition of a fishing zone. It included the UKFIC, British deep-sea fishing interests, conservationists and many islanders.[99] Secondly, the possibility of establishing a multilateral regime was diminished because of Argentina's uncooperative stance.[100]

The colony's prospects were revolutionised from February 1987 by the establishment of the 'Falkland Islands Interim Conservation and Management Zone'. This extended for 150 miles around the Falklands apart from a small quadrant in the south-west. The Falkland government was responsible for the cost of enforcement and the revenues. As a result the income of the Falkland government almost quadrupled within three years, to £28,958,000 in 1987–8. The cost of administering and patrolling the fishing zone in that years was about £6.6 million.[101] The Falkland government recognised that the commitment of the British government would only be sustained in the future if the colony became an economic asset. It therefore pursued an ambitious project to revive the plans of the SAFC for the creation of a British deep-sea fishing industry based in the Falklands. Joint venture arrangements were formed with British and foreign fishing interests which purchased six fishing vessels, at an average cost of £2 million each, as the basis for a fleet. The majority of the ships were refurbished in British shipyards (at Hull and Tyneside) and were manned by British crews.[102] The project was later suspended as a result of the Seamount fiasco, when one of the joint venture companies collapsed with heavy debts.[103]

The Falkland Lobby also participated in the debate about immigration within the colony which preceded the publication of the 'Falkland Islands Development Strategy Final Report' (known as the Prynn Report). The Prynn Report predicted a total population of 2,965 (an increase of 48.3 per cent from the 1986 figure) by 1996–7 if its programme were implemented.[104] A public meeting in Stanley expressed overwhelming agreement that immigration should be restricted to forty people a year.[105] Nevertheless, many in the UKFIC believed that the Prynn Report should have sought to maximise the population for political reasons. In July 1988 Robert Elgood, a member of the UKFIC, warned in a letter to the colony's independent newspaper that:

> Islanders should face the hard fact that only with what appears to be a massive population increase can the Falklands hope to win support from enough U.N. Member States to safeguard their sovereignty . . . Sooner or later a U.K. Government will be elected that for whatever reason, will wish to cut its support for the Islands.[106]

Bill Hunter Christie argued that the desired population figure should be from 10,000 to 20,000.[107]

To conclude, in the mid-1960s a negotiated settlement of the sovereignty dispute with Argentina seemed a logical step in the withdrawal from empire. The Falklands had a precarious economy, could not sustain independence and were isolated by their geographical position. However, the formation of the Emergency Committee and its subsequent campaign led to an assurance from the British government in 1968 that sovereignty would not be transferred against the wishes of the islanders. During the early 1970s the pressure group formulated a vision of how economic development could create a sizable, self-sufficient British community in the Falklands. The 1976 Shackleton Report added to speculation that the Colony had a future independent of Argentina. The activities of FIRADA developed a perception in the minds of many MPs that the Falklands were a valuable economic asset to Britain. This may have influenced their response to the 1980 leaseback initiative and the 1982 war. The Falkland Lobby fulfilled its immediate objective of countering Argentine pressure on the British government to disregard the islanders' wishes on sovereignty. Nevertheless, its influence was largely limited to this single issue. The failure to win support for the implementation of the 1976 Shackleton Report programme would have probably culminated in the colony's economic collapse. It was the Argentine invasion of the islands in 1982 which transformed the dispute and gave the Lobby's ideas a new lease of life.

NOTES

I should like to record my gratitude to everyone who assisted me with my research. I am especially indebted to Bill and Merle Hunter Christie and Frank Mitchell of the Falkland Islands Association, and to my former supervisor John Darwin. I am also grateful to Alex Danchev for his comments on an earlier draft of this chapter.

1. Cmnd. 8787, *Falkland Islands Review* (London: HMSO, 1983) (hereafter Franks Report), p. 6.
2. James Callaghan, *Time and Chance* (London: Collins, 1987), p. 371. The Falkland Islands Company owned about 40 per cent of the Colony's land area until 1980 and dominated the local economy. It enjoyed a virtual monopoly in shipping, banking and retailing.
3. G.M. Dillon, *The Falklands; Politics and War* (London: Macmillan, 1989), pp. 55–89.
4. See C.R. Ellerby, 'British interests in the Falkland Islands: economic development, the Falkland Lobby and the sovereignty dispute, 1945 to 1989', unpublished D Phil thesis (Faculty of Modern History, Oxford University, 1990).
5. *Official Report* (House of Commons unless otherwise stated), vol. 761, cols. 1458–62, 26 March 1968.
6. Franks Report, p. 6.
7. Ibid.
8. F.G. Mitchell's private papers: unofficial members of the Executive Council to all MPs, 27 February 1968.

9. F.G. Mitchell interview, 21 July 1987.
10. *Official Report*, 26 March 1968, vol. 761, cols, 1458–62.
11. Sir Cosmo Haskard to C.R. Ellerby, 25 February 1988; *The Times*, 19 September 1988.
12. C.J. Christie, 'Nationalism and Internationalism: Britain's Left and Policy towards the F.I.s, 1982–1984' (University of Hull, April 1985), Hull Papers in Politics, no. 37, p. 3.
13. E.W. Hunter Christie, *The Antarctic Problem* (London: George Allen and Unwin, 1951).
14. E.W. Hunter Christie interview with Michael Charlton, 'The Little Platoon' (BBC, Radio 3), broadcast 17 May 1987.
15. Mitchell papers: FIs Emergency Committee minutes, 25 March 1968.
16. UKFIC archive: UKFIC minutes book, 20 August 1968.
17. *Daily Telegraph*, 28 November 1968.
18. S.E. Finer, *Anonymous Empire* (London: Pall Mall Press, 1966), p. 93.
19. *Daily Telegraph*, 30 November 1968.
20. UKFIC archive: group C, file 16 (henceforth cited as C(16)), Christie to Prior, 9 December 1968.
21. Ibid.
22. *The Guardian*, 13 December 1968.
23. Andrew Roth, *Heath and Heathmen* (London: Routledge and Kegan Paul, 1972), pp. xvi and 190–205.
24. David Childs, *Britain since 1945*, 2nd ed. (London: Methuen, 1986), p. 201.
25. *The Guardian*, 8 December 1968.
26. Max Hastings and Simon Jenkins, *The Battle for the Falklands* (London: Michael Joseph, 1983), pp. 22–3.
27. FIs Legislative Council (Leg. Co.) minutes, 25 October 1973, *Falkland Islands Gazette.*
28. Falkland Islands Company records (London): Mitchell, 'F.I.s Communications', September 1971.
29. Falkland Islands Association (FIA.) archive (London): 3(E), Memorandum, 'Support for the F.I.s in London', undated.
30. The Reverend P.J. Millam, interview, 9 March 1988.
31. UKFIC archive: B(3), Christie to Barlow, 15 June 1972.
32. C.R. Ellerby's private papers: UKFIC/FIA Memorandum, Constitution and Membership of the UKFIC, 1987.
33. UKFIC archive: FIs Committee, 23 May/26 September 1974.
34. Ibid., 13 May 1974.
35. Ibid., B(9), Christie to Leg.Co. members, 17 July 1974.
36. Ibid., Luxton to Christie, 16 October 1974.
37. Ibid., Christie to Leg.Co. members, 17 July 1974.
38. Ibid., UKFIC minutes, 8 April 1974.
39. Hastings and Jenkins, p. 326.
40. FIA archive (London): 3(B), UKFIC, Memorandum, 'The F.I.s Oil and Natural Gas', April 1975.
41. L.S. Gustafson, *The Sovereignty Dispute over the Falkland (Malvinas) Islands* (New York: Oxford University Press, 1988), p. 89.
42. F.I.A. archive (London): 3(B), D.G. Ainslie, Memorandum, 'The future

of the F.I.s with particular reference to the implications of oil existing under or around the Islands', November 1974.
43. E.W.H. Christie interview, 13 May 1988.
44. FIA archive (London): box file 5, Christie, Report, April 1975.
45. NAI French interview, 12 Dec. 1987.
46. UKFIC archive, D(9), Christie to Hall, 31 December 1974.
47. *The Times*, 21 July 1976.
48. Lord Shackleton (Chairman), *Economic Survey of the Falkland Islands* (London: Economic Intelligence Unit, May 1976), vol. I, passim.
49. *Official Report*, 2 February 1977, vol. 925, col. 551.
50. Ted Rowlands interview with Michael Charlton, *The Little Platoon* (B.B.C. Radio 3), broadcast on 3 May 1987.
51. Shackleton Report (1976), vol. II, 93–9.
52. Rev. P.J. Millam's private papers: UKFIC paper, Christie, *Assessment of the Prospects for the Islands*, undated.
53. Ellerby papers: Chairman's Report, FIRADA, 26 September 1978.
54. FIA archive (London): 5(A), Sallingbury Ltd., *Sovereignty of the F.I.s*, 22 September 1976.
55. *Falkland Islands Newsletter* (London: FIA), 18 June 1978.
56. FIA archive (London): 5(B), *What is the F.I.s Office?*, 1979.
57. Ibid., 5(A), Joint UKFIC/FIRADA Executive Board minutes, 28 February 1979.
58. *Falkland Islands News* (Lincoln's Inn Reception Issue), 19 June 1979.
59. Ellerby papers: Chairman's Report, FIRADA, 9 November 1979.
60. FIA archive (London): 5(B), *What is the F.I.s Office?*, 1979.
61. Ibid., box file FIRADA, FIRADA Executive Board minutes, 17 April 1980.
62. FIs Leg. Co. minutes, 20 June 1979, F.I.s Gazette.
63. Ellerby papers: Chairman's Report, FIRADA, 9 November 1979.
64. Mark Wise, *The Common Fisheries Policy of the European Economic Community* (London: Methuen, 1984), pp. 142–60.
65. Mitchell papers: SAFC minutes, 28 July 1977.
66. FIA archive (London): box file 5, SAFC, 'Proposal for a Commercial Fisheries Survey of the South West Atlantic', April 1978.
67. Mitchell papers: SAFC minutes, Buchanan-Smith to Rowlands, 19 June 1980.
68. Wise, *Common Fisheries Policy*, pp. 47–51.
69. Franks Report, pp. 22–3.
70. *The Falkland Islands Times* (Port Stanley), December 1980.
71. Franks Report, p. 23.
72. *Official Report*, 2 December 1980, vol. 995, cols. 128–31.
73. Franks Report, p. 23.
74. FIA archive (London): 9(B), UKFIC minutes, 17 December 1980.
75. Ibid.
76. P. Watts interview, House of Common Foreign Affairs Committee, Session 1982–3, *Falkland Islands*, Minutes of Evidence (London: HMSO, 4 February 1983), p. 292.
77. Ibid., Evans to Hunt, 30 December 1980, Appendix 16 to the Minutes of Evidence, p. 461.

78. Ibid., P. Watts interview, Minutes of Evidence (4 February 1983), p. 293.
79. *F.I. Times*, January 1981.
80. Franks Report, p. 23; Leg.Co. minutes, 7 January 1981.
81. FIA archive (London): 9(C), Malcolm to Frow, 3 November 1981.
82. Franks Report, pp. 23–9.
83. *F.I. Newsletter*, November 1987.
84. Ibid., 17 June 1982–12 November 1982.
85. Falkland Islands Development Corporation, *Report and Accounts*, 1984, Appendix 1.
86. Ibid., 1988, p. 7.
87. Ellerby papers, UKFIC/FIA Memorandum, Constitution and Membership of the UKFIC.,1987.
88. *F.I. Newsletter*, May 1984.
89. Ibid., November 1982.
90. Author's papers, FIA, Report by the Honorary Auditors, Accounts for the Year Ended 31 March 1987.
91. *F.I. Newsletter*, September 1988.
92. Millam papers: Minutes of joint meeting, FIA Executive Committee and Ad Hoc Committee of the UKFIC, 18 February 1988.
93. *F.I. Newsletter*, December 1988.
94. Fifth Report from the House of Commons Defence Committee, Session 1986–7, *Defence Commitments in the South Atlantic* (London: HMSO, 13 May 1987), p. vi.
95. FIA archive (London): file 30, S.A.C. press release, 'Gallup Poll shows public supports improved relations with Argentina', 5 March 1986.
96. *The Times*, 21 November 1983.
97. B. George/W. Little, 'Options in the Falklands-Malvinas Dispute', South Atlantic Council Occasional Papers, no. 1 (London: April 1985).
98. Fifth Report from the House of Commons Defence Committee, Session 1986–7, *Defence Commitments in the South Atlantic* (13 May 1987), p. 40.
99. *F.I. Newsletter*, August 1985/February 1986.
100. Fifth Report from the House of Commons Defence Committee, Session 1986–7, *Defence Commitments in the South Atlantic* (13 May 1987), p. 41. For an alternative interpretation see Peter Willetts, 'Fishing in the South-West Atlantic', South Atlantic Council Occasional Papers, no. 4 (London: March 1988).
101. Falkland Islands Development Corporation, *Falkland Islands Development Strategy* (London: Environmental Resources Ltd., August 1988), henceforth cited as *Prynn Report*, p. 14.
102. Stanley Fisheries Ltd., *Report and Accounts*, 1987, pp. 6–9.
103. *The Observer*, 11 June 1989.
104. *Prynn Report*, August 1988, passim.
105. *F.I. Newsletter*, September 1988.
106. *Penguin News*, 22 July 1988.
107. Millam papers: Minutes of Joint Meeting of the FIA Executive Committee/Ad Hoc Committee of the UKFIC, 18 February 1988.

5 International and Inter-Agency Misperceptions in the Conflict[1]

Virginia Gamba-Stonehouse

One of the most interesting aspects of Anglo-Argentine conflict over the Falkland Islands is that the parties to the underlying sovereignty dispute had been misleading one another for many years over the intention and timing of their bilateral negotiations. Within that conflict, this chapter focuses on the twin complicating factors of international communication failures and inter-agency communication breakdowns before and during the 1982 war. International communication failures connote those issues which sent the wrong signals to the counterpart in a negotiation; were misinterpreted at the time; or demonstrate that important information was not delivered on time or at all. These failures were aggravated by misperceptions, some of which indicate long–standing prejudgments and/or poor understanding of a reality. Inter-agency communication breakdowns connote those issues which evidence internal communication problems between governmental institutions; rivalries between governmental agencies in general and between armed services in particular; and domestic pressures on the decision-makers generated by extra-governmental organisations. We may observe that it is difficult to separate international communication failures from those which are originated in inter-agency communication breakdowns. More often than not, one set of misperceptions led to the other type of communication breakdowns. Indeed, it is difficult to ascertain to any great extent whether inter-agency and domestic squabbles were the *cause* and not the *consequence* of misperceptions.

Since the Falkland/Malvinas issue is a long-standing territorial dispute it presents three different phases in the developments that led to the 1982 war. Thus, it is simpler to identify both international and inter-agency communication failures at each phase than to see them as a whole. These phases are: conflict, crisis and war.

The conflict phase runs from the very beginning of the dispute between Britain and Argentina, but for practical purposes it is best considered from the mid-1960s when the Falklands question started to be negotiated directly at the bilateral level following the mandates of successive United Nations resolutions. The crisis phase can be said to begin by December 1981 and extend up to 1 April 1982. In this phase the inertia of the dispute was overcome; the Argentine government gave priority to the islands, the bilateral negotiations suffered a set-back and the South Georgia crisis erupted. The last phase is that of the war proper. Here the time frame is from 2 April to 14 June 1982, covering Argentine occupation and British recovery.

There appear to have been a number of significant international communication failures during the conflict phase. The first of these was the tacit message sent by the British Foreign Office to Argentina by not voting against UN Resolution 2065/XX in 1965. There is no doubt that at one time, long ago, the issue of the islands had gained importance at the Foreign Office level in London, but after 1945, British interests in these specific South Atlantic territories declined. When the Wilson administration decided on its policy to withdraw from east of Suez and concentrate on Europe, one of the many places listed for decolonisation by the United Nations was the Falkland/Malvinas Islands. In Argentina the issue was a lot clearer, precisely because the Argentine claims had remained static over a long span of time and were always a matter of high priority in its foreign policy.[2] When in 1964 the British placed the islands on the decolonisation list, the Argentine government immediately mounted a diplomatic offensive in the international arena. In Argentina's perception, decolonisation of the territory presupposed some sort of transition to self-determination for the islanders and eventual territorial autonomy. Argentines felt that the islands had no right to self-determination and could not claim formal independence because the inhabitants were not the original population, which had been displaced by force by the British expedition of 1833.[3]

Argentina went to the only international forum where it felt its side would be heard: the United Nations. The diplomatic offensive was designed to stop Britain from listing the territories under the decolonisation issue. Argentina insisted that the problem was a bilateral Anglo-Argentine problem. Any talk by the British of consulting the wishes of the islanders by means of some kind of referendum brought retorts in kind by the Argentines:

> By way of anecdote, I remember that when Chancellor Callaghan told me in the interview he held with me – then I was Minister of Foreign Affairs for Argentina – that they had to consult the wishes of the islanders, I answered that it was a pity that the British had not consulted the wishes of the islanders when they occupied the islands by force in 1833. He could not answer except with a smile.[4]

The Argentine diplomatic manoeuvre succeeded in the United Nations, as expressed in Resolution 2065/XX of the General Assembly.[5] The resolution urged Argentina and Britain to proceed without delay with negotiations with a view to finding a peaceful solution to the problem. Though the British representatives did not vote against the resolution, they had attempted in the past to have the self-determination of the islanders and their wishes expressed in any resolution from the United Nations. This was not to be and they were effectively outflanked.[6] The British were surprised at how the Falklands issue had developed in that forum; they still preferred to abstain from voting against Resolution 2065/XX, therefore sending a signal to Argentina of renewed hope in the seriousness of London's will to negotiate the end of the dispute through the mechanisms set out by the United Nations.[7] In Argentina, the British abstention and the approach of the Foreign Office over the next two years, during which bilateral negotiations almost culminated in a sovereignty transfer, were received as signs of a British willingness to find a quick and easy solution to the dispute.[8] This issue was inextricably linked to a second: the British attitude to negotiations, and in particular the negotiation of the sovereignty issue.

Until 1939, Argentina's claim to the islands was part of the small change of diplomacy, a matter of postal conventions and arguments over whether the islands and mainland would accept each other's stamps. Nevertheless, the advent of Juan Domingo Peron in Argentina determined a new rush towards the reinforcement of old geopolitical concepts in Argentina.[9] Under his regime, the Argentine government – in 1946 – resubmitted its claim to the islands. One of the effects of the rise in nationalistic attitudes in Argentina was that over the next forty years the Argentine interest in the islands ebbed and flowed with the Peronist cause. It had become an important political factor.

In an effort to counteract Argentina's claims that the islanders were victims of colonial rule, Britain decided to amend the islands' constitution during 1948. This introduced a degree of democracy in the islands which led to the beginning of the trappings of self-government. This was seen in Argentina as a move away from any possible negotiation in future for the sovereignty of the islands. Thus a pattern was gradually established: the Argentines would instinctively react to what they saw as British initiatives to change things in the islands, whether constitutionally or economically. In 1962, for example, in the wake of the successful signing of the Antarctic Treaty (1959) by several countries, Britain decided to create a new organisation in defence of its Antarctic claims. This was deemed necessary at the time since the Treaty of Antarctica demanded the demilitarization of the region and it would entail a loss of British presence in the territory. To fill the void, Britain created the BAT (British Antarctic Territories), a precursor of the BAS (British Antarctic Survey). This was regarded in both Argentina and Chile as a breach of faith in the Antarctic Treaty. It was then that Argentina revived her claims to the Falklands in the United Nations.

By 1968 the *status quo* was changed through pressure from within the British government. The formation of the Falkland Lobby, and their attack on the transfer of sovereignty, forced the Foreign Office to change the parameters of discussion and to ask the Argentines once again for the freeze of sovereignty. The Argentines understood that the British government was under attack at home, and they bent over backwards to comply with the request. This led them to accept the 1971 Communication Accords which reduced the British expenses of maintaining its own lines of communication with the islands, making them much more dependent on Argentina than ever before. Even then, the talks continued and finally sovereignty was once more placed as part of the original negotiating package.

When this was taking place, the Foreign Office sent Lord

Shackleton in 1975 to make a review of the economic possibilities that the islands offered for self-support in the future. This move again woke in the Argentines their old fears of self-determination in the islands. This was aggravated by the simultaneous naval incident relating to the *RSS Shackleton*. This Royal Survey Ship was shot at by an Argentine navy vessel when it navigated in waters that Argentina claimed as her own. But the Shackleton Mission appeared to be the real threat. The Argentines felt that the British had negotiated with them for a number of years, but were not prepared to go all the way and thus were initiating a strategy by means of which the islands would be quietly helped to self-determination and economic self-reliance.[10] The incident broke the relations between both countries until 1977–8, when again the British agreed to bring back the sovereignty issue to the package under negotiation, which led to the leaseback proposal of 1980.

When the leaseback proposal was shouted down in the House of Commons in December of that year, the Argentines perceived it as yet another procrastinating manoeuvre. The episode in the Commons led not only to the proposal to freeze the sovereignty issue once again but also to the demand to add a member of the Falkland Islands Council as party to all further bilateral negotiations. The Argentines felt that the situation had considerably worsened. The negotiation mechanism was being changed to add a third party to the dispute – a move that, if agreed by Argentina, might give grounds for the self-determination of the islanders in future. After the diplomatic protests which ensued, relations between both countries were again mended and a date accepted to renew negotiations by December 1981 at a bilateral level. Nevertheless, by October 1981, new elections in the islands prompted the entrance upon the scene of extremely pro-British council members in Stanley who lost no time in demanding from London the assurance that no sovereignty would ever be negotiated with the Argentines.[11] This was the status of the negotiations when the new military junta took over in December 1981. It is not surprising that they immediately arrived at the conclusion that the British Foreign Office was not serious or constant in the negotiations. Buenos Aires had received many contradictory signals which pushed them to the belief that Britain refused to understand the seriousness with which Argentina viewed this issue. The junta understood then that they had no power to make sure that the British kept to a continuous pattern of bilateral negotiating terms.[12] It was then that they toyed with the idea of using force to extract attention from London.

Argentina might not have been so confused about the signals she received from London during the negotiations if the British commitment to defend the islands had been demonstrated less ambiguously. But clarity on the commitment to the islands was not forthcoming. On all the previous incidents with military overtones, Britain had reacted in similar terms: diplomatic protestations and ineffectual action for the defence of the islands. The military garrison itself had only been upgraded from one to forty men as a result of an incident in 1966. No additional naval assets had been assigned to the area. On the contrary, the Foreign Office urged Argentina to consider the establishment of all direct communications between the islands and the mainland, despite the fact that sovereignty had temporarily been removed from the agenda. The Argentines saw the advantages pointed out to them by the Foreign Office: get to know the islanders and they will get to trust you. Some day the the islanders themselves might wish be to be Argentine.

When the Argentines reaffirmed their rights over southern Thule, for example, by establishing a scientific research station there in late 1976, the British government reacted with no more than a formal protest. It took two years for the issue to hit the British press and even then, it was relegated to the bottom of the page.[13] We know now that the British decided to make a covert naval deployment to the South Atlantic in November 1977 – an episode discussed in Chapter 6 of this book – because of the 'dangerous' posture the new Argentine military government was adopting regarding the islands, yet this gesture never reached the Argentines at all.[14] Almost immediately the Foreign Office entered the most far-reaching and revolutionary negotiation phase with Argentina since 1967. This eventually led to an understanding on a leaseback agreement by 1980.

Thus the Argentines were convinced that the British were willing to start meaningful talks again because they thought that London had appreciated the resolution of the new Buenos Aires government vis-a-vis the South Atlantic. As far as the Argentines were concerned, London had not reacted strongly to any of the Argentine moves since 1977. As far as the British were concerned, they thought that the Argentines had abandoned the idea of possibly resorting to force because they had been deterred by the sending of a task force. It is interesting to note that the British were convinced that the Argentines had been aware of the manoeuvre because they had particularly asked the Americans to convey the message to Buenos Aires. This belief that the Americans could serve as an efficient communication line also impinged on decisions during the 1982 war. The reality was that the Americans passed no such message to Buenos Aires at the time. Nonetheless, even under the misperception that the Argentines had been aware of the task force, they agreed to negotiate on better terms (for the Argentines) than ever before except for 1967.

Finally, by 1981, even after the Houses of Parliament had rejected the leaseback proposal, nothing was done to reinforce the islands militarily in the expectation of an Argentine reaction. In fact, the contrary occurred. Notoriously, in June 1981, the last Royal Navy patrol ship to serve the islands, *HMS Endurance*, was withdrawn from service, a decision to be effective by April 1982. The British Antarctic Survey base in South Georgia (Grytviken) was threatened with extinction through lack of funds. These were hardly the messages to be sent to Argentina precisely at a moment when the diplomatic initiatives for leaseback had floundered.

Fourthly, there was the aggravating factor of the misperceptions in Argentina relating to the possibility of a Chilean-British collusion in the area of the southern Atlantic and sub-Antarctic waters – an issue examined from the Chilean perspective by Felipe Sanfuentes in Chapter 3. The rivalry between Argentina and Chile over the territories of Patagonia, Tierra del Fuego and Antarctica is a long-standing problem.[15] By 1881 Argentina and Chile's gravest dispute was over the possession and control of the Magellan Strait. The signing of a treaty to keep the Strait demilitarised solved the problem and no further border conflicts resulted until the 1960s when new claims on the territories south of the Beagle Channel were presented by Chile. This led to an agreement to submit the issue to arbitration by the British Crown and, by 1975, the arbitration was ruled in Chile's favour. Subsequently, the Argentines complained about the ruling and demanded further negotiations. The issue became so critical that Argentina and Chile almost went to war in December 1978. By January 1979, both countries had accepted the Vatican's mediation to rule again on the territorial issue, thus averting war. In 1980 the Pope ruled in favour of Chile and this led to further complaints from Argentina who persuaded both the Vatican and Chile to continue the deliberations.

What is important in this issue is that, in the minds of most Argentine nationalists and of the military, the territorial disputes that Argentina held with Britain over the Falklands, with Chile over the Beagle Channel and with both over sovereignty in Antarctica, were connected. Since the British Crown had been the arbiter on the Beagle Channel dispute and had in 1975 ruled in favour of Chile, a link was made in Argentine perceptions relating to the ruling of 1975 and the initiation of the Shackleton Mission in the Falklands in the same year. Since the Shackleton Mission led to a serious breakdown of bilateral relations and of the ongoing negotiations, all of these were taken as part of a whole:

> The Government of her Britannic Majesty, by ruling in favour of Chile on the Beagle Channel issue has sought not only the Chilean presence in the South Atlantic but also to place her in the southern flank of the Malvinas islands. In this way, Britain has stimulated Chile to interfere in the conflict (the Falkland/Malvinas Islands conflict) and this has started to manifest itself already.[16]

This linkage concept was exacerbated by two factors: the changes occurring as of the 1960s relating to the extension of territorial waters and economic zones in the sea in South America, and the renewed emphasis on the Antarctic territorial claims by the countries concerned.[17]

After the installation of military government in Argentina in 1976, many voices were raised over the need to reaffirm Argentine rights over the South Atlantic and Antarctic regions. Most academics at the time believed that the future of Argentine lay in the efficient control and exploitation of the maritime expanses facing Patagonia and south towards Antarctica. Of these regions, two countries limited further Argentine involvement: Chile because it held contrasting claims that, if successful, would allow her entry to the South Atlantic (via the enforcement of the 200-mile ocean limit of the islands south of the Beagle Channel); and Britain, because she administered a territory claimed by Argentina and over which Argentines had no right to possess land or bases. It was the common belief in Argentina that the situation could be controlled as long as the two competitors in the region were isolated one from the other. As long as the situation related to the arbitration and mediation with Chile remained unresolved, the problem of the Falklands was a mere irritant at Anglo-Argentine relations level. However, should the Chilean claim prosper, then progress had to be made with Britain over the Falklands issue. The rationale was that the successful Chilean claim in the south might cut off the Argentine projection to Antarctica and make it necessary to base her own power projection from the islands in the South Atlantic rather than from the territory of Tierra del Fuego.

It is obvious that the worst possible scenario that many Argentines saw at the time was the possibility of an Anglo-Chilean collusion of interests in the South Atlantic. In theory this would have been possible given two outcomes: Chile gaining her claim, bringing her very near the limit of the surrounding waters of the Falkland Islands Economic Zone and territorial waters; and Britain deciding to freeze all sovereignty talks with Argentina, risking being cut off from the continent by an irate Argentina and thus being free to seek alternative connections for the islands with the new-found Chilean position in the South Atlantic. This worst-case analysis was what had rocked Argentine decision makers to the core when in December 1980 not only the Pope ruled in favour of Chile, but the British Parliament vetoed a leaseback agreement on the islands with the Argentine. The eyes of the Argentine military decision-making body become glued on the South Atlantic region.[18]

It is surprising that the parties to bilateral negotiations in 1978 should not have given more importance to the way this third issue was impinging in the future of the negotiations. After the debacle in the Commons in December 1980, the British Foreign Office seems not to have attached much importance as to how this issue, coupled with the Vatican's ruling in favour of Chile in the South Atlantic, was creating pressure on Argentina and forcing the decision-makers to take up much harder stands than before. Later in 1981 Lord Carrington's surprising decision to 'talk for the sake of talking' with the

Argentines – rather than to promote lease-back – demonstrated an equal lack of understanding of how this problem had become linked with other territorial concerns in the South Atlantic region. The British intelligence assessment that Argentina posed no threat towards the end of 1981 seems to have been based on wholly deficient analysis or increasingly out-of-date information – a consequence, perhaps, of a progressive reduction in the British capability for information-gathering in the region as a whole.[19]

Turning to inter-agency and/or internal communication problems which originated in the course of the negotiations, two in particular deserve to be highlighted: in Britain, as a consequence of the creation of the Falklands Lobby; and in Argentina, as a consequence of the internal changes in force composition and their impact on Argentine foreign policy in the 1970s. The creation and activities of the Falklands Lobby are examined in detail by Clive Ellerby in Chapter 4. By its very nature the Lobby exploited all the information networks it could establish within government and in the media. This led to the exacerbation of inter-agency (in this case inter-ministerial) problems affecting the Falklands. In 1968, for example, it was a leak from within the Ministry of Defence which gave the Lobby the information they needed to attack the secret negotiations the Foreign Office was conducting with the Argentine government over the future of the islands. British invocation of prior agreement by the islanders to any transfer of sovereignty did not emerge as a basis for bilateral negotiations until after the emergence of the Lobby. Cornered by the disclosure of the secret negotiations, the government had to relent by stating that from then on they would take into account the wishes and not the interests of the island population before any transference was discussed. It is not surprising that Argentina resented not only the Lobby's existence, but its success.[20]

Important examples of the Lobby's strategy relate to the effective use of the media in London to generate awareness of the issue and the effective use of the Houses of Parliament to demonstrate this opposition. Many instances exist of both cases, such as the press leaks in 1968, and – more importantly – in March 1982 which were directly related to the South Georgia crisis. Parliament played a major role in the escalation of the conflict and the crisis in such instances as the 1968 debates, the

1980 rubbishing of the leaseback proposals, the December 1981 discussion of the future of British presence in Antarctica and in the islands (timed so as to be coincident with the initiation of bilateral talks between Britain and Argentina on the future of the islands), and finally in the Commons debates on 23 March 1982. In the last case, it seems likely that the tenor of the debate and the motion presented the following day by some 100 MPs asking the government to defend the islands in future and abandon all sovereignty negotiations with the Argentines was a significant influence on the Argentine decision of 26 March to intervene militarily in the Falklands by 2 April.[21]

Another case of inter-agency communication problems concerns the changes in Argentine force structure and their impact on Argentine foreign policy in the 1970s. By the end of the 1960s the domestic and economic situation of Argentina had deteriorated. Inter-service rivalries had originated bitter fights to obtain influence, power and substantial budgets from the central administration. On top of this the beginning of insurgency in the country had started to generate disorder and insecurity among the population. The upheaval grew to extraordinary proportions and permitted the return of President Peron from exile. By 1974 the Perons were ruling the country. Nevertheless, the insurgency movements that had brought about the change continued to make demands on the administration despite the return of democracy. Peron refused to concede more political power to the extremist groups and a year later passed away leaving his wife, Isabel, to rule the country. By 1975, the Peronist government was besieged by the demands and the terror caused by open rural and urban guerrillas. It responded by commanding the military to 'annihilate' the problem and also by creating its own paramilitary elites of the extreme right. What followed was the beginning of the Argentine Repression. It took little time for the military to conclude that if they were made to bear the brunt of repression, they should also govern. From 1976 to 1980 the military proceeded vigorously to annihilate subversion. Other priorities of the government were to restore economic stability, and to obtain good regional and international alliances that would support the junta. Not surprisingly, the government consisted almost in its entirety of military officers acting in Cabinet posts and as intervenors in state enterprises. Policies at all levels of government started to suffer from a militarisation process where the value systems of the military were superimposed on those of civilian society. With the return of the military came the return of the old concept of territorial dominance in the determination of national interests.[22] This was partly a result of a new interest to keep the borders under control and partly as a result of isolation due to the nature of the government.[23] Almost all the policy prescriptions of the Peronist government were changed. Nevertheless there was one issue that all parties agreed to treat in like fashion: the Malvinas.

In 1976, the priorities in the field of foreign policy were the territorial disputes with Chile, Brazil and the United Kingdom. Seeking to reassert their territorial presence, the military pushed Chile to the point of war by December 1978.[24] On the Falklands issue the situation was tense in 1976 and 1977. It was only by 1978 and 1979 that serious negotiations at bilateral level started. With the leaseback proposal sponsored by the ill-fated Nicholas Ridley, the relations between the Argentine military and the Thatcher government improved. By 1980 the possibility of escalating the territorial dispute with Britain began to disappear. However, in December 1980, with the defeat of the leaseback proposal in the House of Commons and the hardened posture of the islanders, came a period of uncertainty. In these circumstances, given the changing policy values and the militarisation of Argentine society, it was impossible to avoid strong inter-agency and inter-service problems from occurring. Some of these influenced the decision-makers who formed part of the military junta that led the country to war. Among these problems the following seem to stand out in the influence they held on the Malvinas issue: the rivalry between the three armed forces, the power partition at the level of the three-man military junta on regional responsibilities, and intra-bureaucratic communication problems and rivalries at ministerial and service level.[25]

Lacking a centralized command structure as well as a strong civil service bureaucratic structure, Argentina has often suffered from inter-service and inter-ministerial rivalries. The governing system in Argentina does not encourage input from the bottom in the ministerial structure. More often than not, with each change of government the political appointees to the top positions of the ministries also change. Communications are severed from this moment on: the top elites organise the policy and implementation of policy through a much reduced nucleus of individuals, none of whom belongs to the bureaucracy of the relevant ministry. Two levels of action become evident – the topmost structure of political appointees with their own personal advisors, and the residue of routine duties which fall to the entrenched bureaucrats. This effectively ruptures communication mechanisms at ministerial level.

Though one would have assumed that the Malvinas issue was in the domain of the Ministry of Defence, this was often not the case. With a ruling military junta made up of the three armed forces' commanders-in-chief, the services controlled power directly and did not need a strong Ministry of Defence. Thus, the retaining of power by the three armed services at the top generated a purposeful undermining of the power and status of two military institutions: the Ministry of Defence and the Joint Chiefs of Staff. The weakness of these institutions was also the cause of the other ministries' underestimation of the Ministry of Defence. The net result of this system structure was that, at a time of the 1982 war, the mechanism for co-ordinating military and diplomatic actions at inter-agency level was non-existent.

During the years where the Falklands conflict gathered impetus, the main responsibility for the follow-up of the Falklands issue lay with the Ministry of Foreign Affairs and in the individual hands of the three armed services. These too were seldom made compatible. The problem lay in the *type* of governmental system that Argentina experienced at the time. The difference between the Argentine military government as of 1976 and, say, that of Chile was that the former was the joint rule of the armed forces whereas the latter was a one-man military dictatorship. This meant that the Argentine armed forces – represented by their commanders-in-chief – ruled the country as a junta where all responsibility for the administration and security of the country were equally shared.[26] Argentina was literally divided up between the three services. In this fashion, Patagonia, Tierra del Fuego and the territories in the South Atlantic and Antarctica were the main responsibility of the navy. Naturally where such a rigid partition system was established, competition and jealousies were rampant. Each service was constantly striving to better its position at the expense of the other two.

The competition for power and money that the three services

initiated under the junta structure caused severe problems for inter-service dealings.[27] This was also a result of the profound insecurity that the country was experiencing in its war against internal subversion. Insecurity and guerrilla tactics inevitably led to the adoption of a cell-like counter-insurgency structure. Copying guerrilla tactics, the military formed its own groups which acted swiftly, in close co-ordination and in isolation from the main line of the military activities within their service. Co-ordination was only achieved at the very top, that is, the junta itself. But counter-insurgency intelligence, missions and forces were manned and planned by each individual service. This system, though ultimately effective for the purposes created, was a negative influence on the overall centralisation of administration and responsibilities within the government. The result was a fear of co-operation and co-ordinated policies and actions. Such a system caused tremendous military and political problems during the Falklands war.

The main responsibility for what occurred in the south of Argentina, the Antarctic Sea and Antarctica was given to the Argentine Navy. Because of this, the navy was overzealous in the determination of its rights and responsibilities in this region. The two most pressing territorial disputes at the time were in the region assigned to the navy and revolved around the possession of islands and the projection to Antarctica. In these, the navy's chief competitor was the Argentine air force. The navy attempted to project its influence not only in those issues that had to do with navigation and transport in the sub-region, but also in the field of geopolitical aspirations. In this sense, the commanders-in-chief of the navy, especially since the advent of Admiral Massera (who was one of the three members of the junta in 1976) ordered numerous war-gaming and training exercises to project Argentine naval power south against Chile and against the Falkland Islands and their Dependencies. These decisions were not of direct concern to the other two junta members and often they were not immediately informed of such initiatives. Thus it is difficult to ascertain who exactly was responsible for the decision to establish the research station on Southern Thule in 1976–7.

By the same token, the much-discussed possibility of creating such a station on South Georgia – the famous Project Alpha – was also not a clear-cut decision of the junta's. Again it seems that it was more a naval decision dating from before the constitution of the Fourth Military Junta in 1981. When the Galtieri Junta took over in December 1981 and the situation with Britain deteriorated, these cases of naval operations and initiatives in the South Atlantic came to the attention not only of the commander-in-chief of the navy but also of the junta as a whole, and were therefore discussed at that level. This explains the junta decision in January 1982 to postpone Project Alpha indefinitely. The Minister of Foreign Affairs, at the time Dr Nicanor Costa Mendez, was also able to use this information at the highest level to recommend that the navy exert caution in its routine dealings in the South Atlantic. This advice appears to have been heeded at the time.

Despite these few cases when an internal policy or decision by a service was actually discussed at junta level, more often than not, departmental decisions were taken in isolation and without ultimate junta approval. Sometimes not even the commander-in-chief of a particular service was aware of routine policies implemented from below. This is understandable since the commander-in-chief of any one of the armed forces in Argentina as of 1976 was more preoccupied with the political mission of administering government than with the military and operational guidance of his force. This role was given to the deputy commander at each service level, who for all intents and purposes had the role that the commander-in-chief should have had in times of normal government. Not surprisingly, not all heads of departments saw it in their interests to keep the topmost echelon totally informed of their actions and procedures. One such issue was the offer to transport Argentine workers to South Georgia at given times of the year. To make matters worse, individual captains of Argentine ships, in the belief that the naval role was to reinforce Argentine rights in the South Atlantic, sometimes ignored navigation codes or the simple acquiescence in the Communication Accords signed with the British. Their excuse was that the British did not always adhere to them either. Hence, it is possible to explain the radio silence kept by the *ARA Almirante Irizar* when it approached South Georgia in December 1981 and later, the reason why the *ARA Bahia Buen Suceso* had not approached Grytviken first during its voyage to South Georgia in March 1982. The problem caused by both incidents was that, because the Argentine naval hierarchy and the junta (and Ministry of Foreign Affairs) were unaware of this behavioural pattern, the British reaction was found both surprising and excessive.

Finally, it must be noted that the civilian population and above all the political parties during the period of military government, especially after 1976, resented the monopoly of political power and the acts of repression of the military. In the first years of terrorism and repression, the political parties and the civilians at large chose to give a blank cheque to the military so that internal order could be restored. Once this was achieved by 1980, pressure grew to make the military hand over the government to civilians. One of the most effective criticisms at this time was that the military had abandoned their natural duty, which was to defend sovereignty. The criticism related especially to the approaching 150th anniversary of the British occupation of the islands. Why did the military not take forceful action against Britain when they seemed to have no qualms about using force against their own people? This criticism mounted and reached a peak in March 1982 during the South Georgia crisis. This daring of the military to show their real nationalism against an external enemy and not an internal enemy did have some effect and influence at the time of the decision to repossess the Malvinas.

We therefore see that international misperceptions and inter-agency disputes and communication breakdowns in the conflict phase of the Falklands war aggravated the overall Anglo-Argentine relationship, hardened all negotiations on the Falkland/Malvinas issue, and led to a flow of information within each country that was partial and self-interested and thus gave little room for constructive and well-informed crisis management during the early months of 1982.

NOTES

1. This chapter is based on research by the author on the subject of 'International and Inter-Agency Communication Failures in the period previous to and during the Falkland/Malvinas War of 1982 between the United Kingdom and Argentina'. This research was conducted during

1989–90 and was sponsored by the United States Institute of Peace, Washington, DC, and supervised by the Department of War Studies, Kings College, London.

2. See Julius Goebel, *The Struggle for the Falkland Islands* (New Haven: Yale, 1982); Adrian Hope, 'Sovereignty and Decolonization of the Falkland/ Malvinas Islands' *Boston Comparative Law Review* (1983); Consejo Argentino para las Relaciones Internacionales (CARI), *Malvinas, Georgias del Sur y Sandwich del Sur: Diplomacia Argentina en Naciones Unidas desde 1945 hasta 1981* (Buenos Aires, 1983).
3. Leopoldo H. Tettamanti, *Los Elementos Objectivos del Caso de las Islas Malvinas desde el punto de vista Historico, Politico y Juridico* (Buenos Aires, 1982).
4. See Manuel Arauz Castex (ex-Minister for Foreign Affairs of Argentina in 1975–6) 'Aspectos Juridicos de la Controversia sobre los Archipielagos del Atlantico Sur', paper presented to the 'Workshop on the Use of Force in the Solution of International Controversies: The Malvinas Islands Conflict', Instittuto Universitario di Studi Europei, Turin, Italy (6 May 1983), p. 29.
5. See Joan Pearce, 'The Falkland Islands Negotiations 1965–1982', in *The Falklands Dispute: International Dimensions* (London: Royal Institute of International Affairs, 1982) pp. 1–4.
6. See Kennedy School of Government, Harvard University, case study on 'Falkland/Malvinas: Breakdown of Negotiations' (1986). Draft by John Lippincott, revised by Gregory Treverton, p. 2.
7. Martin Walker, 'The Give-Away Years', *Guardian*, 19 June 1982.
8. Cmnd 8787, *Falkland Island Review* (London: HMSO, 1983) [hereafter the *Franks Report*].
9. Geopolitics represents the study of the influence of such factors as geography, economics and demography on the politics and especially the foreign policy of a state. In South America geopolitical traditions have been strongly influenced by the late-nineteenth century Prussian school of thought which linked the manifest destiny of nations with their utilisation of space.
10. Castex, p. 14.
11. *Franks Report*, paras 97–9, 106.
12. D. Ruben Moro, *Historia el Conflicto del Atlantica Sur: La Guerra Inaudita* (Buenos Aires, 1985) p. 28.
13. Michael P. Socarras, 'Incident', *Yale Journal of International Law* 10 (1985) p. 368; *Franks Report*, paras 52–7.
14. *Franks Report*, paras. 65–6. Cf. Alex Danchev, 'Life and Death in the South Atlantic', *Review of International Studies* 17 (1991).
15. See Osiris Villegas, *El Conflicto con Chile en la Region Austral* (Buenos Aires: Editorial Pleamar, 1978).
16. Osiris Villegas, 'Las Malvinas: Pleito Improrragable', *Geopolitica* 24 (1982) p. 53.
17. Maria del Carmen Llaver 'La incidencia del conflicto Malvinas en el subsistema del Atlantico Sur', *Geopolitica*, 28 (1984) p. 37.
18. There is a large and significant Argentine literature on the nation's wish to control the South Atlantic. Among these, see: F.A. Milia et al.,

La Atlantartida: Un Espacio Geopolitico (Buenos Aires: Pleamar, 1978); J.A. Fraga, *El Mar y la Antartida en la Geopolitica Argentina* (Buenos Aires: Instituto de Publicaciones Navales,1980); Osiris Villegas, *El Conflicto con Chile en la Region Austral* (Buenos Aires: Editorial Pleamar, 1978).

19. Lawrence Freedman, 'Intelligence Operations in the Falklands', *Intelligence and National Security* (1986) p. 313.
20. David Brown, *The Genuine Seeds of Hostility and War: the Falkland Islands 1816–1982* April-July 1988, pp. 11.
21. Lawrence Freedman and Virginia Gamba-Stonehouse, *Signals of War* (London: Faber, 1990) pp. 68–9.
22. See F. A. Milia et al, *La Atlantartida: Un espacio Geopolitico* (Buenos Aires: Pleamar, 1978).
23. See Abraham Lowenthal and J. Samuel Fitch (eds) *Armies and Politics in Latin America* (New York: Holmes and Meier, 1986).
24. Osiris Villegas, *La Propuesta Pontificia y el Espacio Nacional Comprometido* (Buenos Aires: Pleamar, 1982) pp. 27–37.
25. The views expressed in the next pages on the analysis of interagency problems in Argentina are the product of the author's practical experience working as a civilian analyst for defence within the Argentine government between 1983 and 1986.
26. See F.A.Milia, 'La militaridad argentina en crisis', *Boletin del Centro Naval* (Argentina) 102 (1984).
27. See Virginia Gamba, *Estrategia, Intervencion y Crisis* (Buenos Aires: Editorial Sudamericana, 1985) chs 1 and 3.

6 The Franks Report: a Chronicle of Unripe Time

Alex Danchev

The Principle of Unripe Time is that people should not do at the present moment what they think right at that moment, because the moment at which they think it right has not yet arrived.[1]

* * *

Argentina invaded the Falkland Islands on Friday 2 April 1982. The following Monday the much-admired Foreign Secretary, Lord Carrington, resigned. His ministerial team went with him. Carrington himself spoke resonantly of 'a great national humiliation'. Later he wrote: 'The nation feels that there has been a disgrace. Someone must have been to blame. The disgrace must be purged. The person to purge it should be the minister in charge.'[2] Would this be sufficient expiation? Would Carrington be the first or the last to go? No one could foretell, but as a hastily assembled naval Task Force inched implausibly out of Portsmouth it was hard to be optimistic. Within the week, half-buried in the accumulating pile of political debris, there was the promise of a truly eschatological purgative. On 8 April, in a written answer to a Parliamentary question, the Prime Minister declared her belief that 'there should be a review of the way in which the government departments concerned discharged their responsibilities in the period leading up to the Argentinian invasion'. She would consider what form the review might take and in due course make a statement to the House of Commons.[3]

The promise was made good some three weeks after the Argentine surrender at Port Stanley on 14 June. In this miraculously transformed environment the Prime Minister announced the appointment of a Committee of Privy Counsellors under the chairmanship of Lord Franks, with far-reaching terms of reference:

> To review the way in which the responsibilities of Government in relation to the Falkland Islands and their Dependencies were discharged in the period leading up to the Argentine invasion of the Falkland Islands on 2 April 1982, taking into account all such factors in previous years as are relevant; and to report.[4]

This Committee started work on 26 July 1982. Five brief months later, on an uncommonly auspicious New Year's Eve, their findings were handed personally to the waiting Margaret Thatcher. How the heart must have raced at the sweet, secret moment of discovery! Tensely anticipated in every quarter, the *Falkland Islands Review* (popularly known as the Franks Report) was released on 18 January 1983, instantly appropriated, hotly debated, and as quickly forgotten.[5] Already an anachronism when it began, by the time of its first anniversary the Falklands war seemed to have passed into ancient history. With remarkable speed the Franks Report induced a profound sense of closure in the British body politic, if not the Argentinian. Franks exposed misdemeanours aplenty, but no high crimes. For those inclined to impeach, the sinking of the *Belgrano* remained the only hope, increasingly forlorn.

And yet the Report was, and is, a revelation. Professor W.J.M. Mackenzie's marvellous 'translation' of the Plowden Report on the control of public expenditure applies with equal force to Franks on the Falklands – 'it turns out that the real problem is about the nature of government in general, and of British government in particular. This is what we are discussing, but of course we have to wrap it up in Mandarin prose.'[6] Tony Benn was not joking when he said in the Commons debate that he knew more now about certain developments in the Labour Cabinet of which he was a member than he did at the time.[7] Unseasonably, the Franks Report was a tribune for open government. It afforded detailed and punctual public scrutiny of diplomatic, military and, most notable of all, intelligence traffic between clearly identified agencies and precisely attributed individuals.[8] No British official publication this century has disclosed so much so soon of that forbidden realm where the security classification is king. It is evident that the rigid prohibitions of conventional discourse were carefully but deliberately relaxed. Franks makes an exceptional feast, 'for your eyes only'.

The nature, purpose and effect of the whole exercise became entangled in the grab for immediate advantage from it. Franks and his colleagues found themselves in the unusual position of reviewing a putative diplomatic failure in the light of a decisive military success. Franks himself was reminded of the congressional investigation into the Japanese attack on Pearl Harbor. No proper British precedents suggest themselves, in spite of the Prime Minister's contention that committees of Privy Counsellors had been used before 'to look into matters where the functioning of the government has been called in question and sensitive information and issues are involved'.[9] Perhaps the only possible comparison is with the Dardanelles Commission of 1916–17, established under the chairmanship of Lord Cromer (Evelyn Baring) 'to inquire into the origin, inception, and conduct of operations in the Dardanelles' in 1915.[10] Yet although there were certain similarities to the Falklands review in terms of scope and substance and high political interest, there were important differences between the two. Most fundamentally, the Dardanelles campaign was universally regarded as a disaster – the very reason for the inquiry. As Winston Churchill wrote at the time:

> It seems to me very necessary for the Commission to bear in mind the circumstances in which their inquiry is pursued. The enterprise has ended in defeat and failure. The Army has been withdrawn. The positions which they had won by so much effort and sacrifice have been yielded to the enemy. The hopes, the legitimate expectations, the chances of battle have vanished away; only the slaughter, the suffering and the waste remains. . . .
>
> A great volume of prejudice and not unnatural vexation has gathered round the story of the expedition. All tongues are freed. The natural tendency of the Commission is to look for faults and errors and it is not surprising that they should find them.[11]

The Falklands Committee were often to reflect on how different it would have been if that enterprise too had ended in defeat and failure; but they were permitted the unaccustomed luxury of keeping their reflections to themselves.[12]

The Dardanelles Commission were offered an obvious, indeed notorious, ministerial target. Churchill, as First Lord of the Admiralty and hyperactive member of the Dardanelles Committee, the war cabinet of its day, was widely held to be the ardent and characteristically unrepentant sponsor of the whole operation. Faced with this temptation the commissioners exercised considerable restraint, as the target himself acknowledged in welcoming their Report as 'at any rate, an instalment of fair play'.

> They have swept away directly, or by implication, many serious and reckless charges which have passed current broadcast throughout the land during the long months of the last two years. They have reduced these charges within the limits of modest and sober criticism, and, further, by laying before the nation the general outlines of the story – a long, tangled and complicated story – they have limited the responsibilities which have been thrown on me and under which I have greatly suffered.[13]

Pre-empted by Carrington's resignation, the Falklands Committee were not led into any such temptation – not even by the biggest target of them all, 'she who must be obeyed'. The two functions mentioned by Churchill, however, found a distinct echo in the later Report. Franks went so far as to include as a separate annex itemised 'Comments on some specific assertions', expressly to 'clear up damaging misunderstandings'. For instance, the interesting Assertion that 'clear warnings of the invasion from American intelligence sources were circulating more than a week beforehand' was met by the equally interesting Comment that 'no intelligence about the invasion was received from American sources, before it took place, by satellite or otherwise'.[14]

Franks too laid a story before the nation. Some seventy pages, the first three chapters, of a ninety-page Report consist of an historical analysis beginning in 1965 and becoming progressively more detailed as it approaches the invasion date, culminating in an often hour-by-hour treatment of the last days (paragraphs 15–259). A fourth and final chapter addresses questions prompted by the central issue of the Committee's remit, namely the way in which government responsibilities were discharged (paragraphs 260–339). The three historical chapters have been criticised, on the one hand, for beginning as late as 1965, and on the other for relegating Franks's supposedly more important task – usually construed as sitting in judgement – to the Report's conclusion. It is true that there is barely any acknowledgement of what might be called the prehistory of the invasion; this seems entirely understandable given the tenor of the terms of reference, the degree of urgency imparted by the Prime Minister's eager hope and the chairman's preference for the work to be concluded within six months, and the evident desirability of keeping the final product to manageable proportions for a potentially wide readership.[15] In fact the story that Franks does tell, so often skimmed as a kind of extended prologue to the final chapter (in extremity, the final paragraph), is absolutely central to an understanding of the Report and the strange phenomena it describes. The significance of this story lies first of all in the irretrievable fact of its telling – in itself a prime function of the review. 'The Committee had to be set up,' *The Times* commented, 'if only to provide some objective commentary on a national drama which had brought the country to a pitch of emotional intensity unwitnessed for twenty-five years.'[16] More than this, Franks had to choose what kind of story to relate. The character of his choice is the fascination of the Report.

Comparison with the Dardanelles Commission is especially instructive in matters of personnel. The chairman, Lord Cromer, expired in the middle of the proceedings in January 1917; at seventy-five, he was two years younger than Lord Franks, who remained in exceeding good health throughout. The secretary, E. Grimwood Mears, a former barrister, accepted the appointment on condition that he was knighted for his services, which request was immediately granted. Anthony Rawsthorne, an experienced assistant secretary in the Home Office – a department relatively untainted by Falklands affairs – was well-versed in committee work and positively vetted; delighted with the appointment, he made no such stipulation. There were eight Dardanelles commissioners (including the chairman), a motley group of varying distinction.[17] Not all were Privy Counsellors, whereas this was in effect made a requirement for the Falklands six, one of whom (Sir Patrick Nairne) was sworn in especially for the purpose. The Privy Counsellor is bound by a mighty oath of secrecy and truth. He may be entrusted with the most intimate secrets of state. He may see documents of the highest security classification. He may cross-question prime ministers past and present. What he may not do, in Lord Franks's choice phrase, is blab.[18]

It is now clear that the Committee of Privy Counsellors represented the outcome of a whole series of calculations of the utmost political delicacy. The primary one was its very existence. 'As the Prime Minister probably recognises,' remarked Roy Jenkins, then leader of the Social Democrats, in the Commons debate on the appointment of the Committee, 'an inquiry has always been part of the bargain. It was understood that there should not be criticism during the operation while British lives were at risk, but that she would agree to the fullest inquiry as soon as the operation was concluded.'[19] *Prima facie* evidence for such a bargain is provided by the advance commitment to a review, in many ways a surprising commitment made 'in the heat of the moment to re-establish the authority of a government at war', as the *Economist* underlined. In this sense, perhaps, the government was 'forced' to make a concession, as was persistently alleged by its critics.[20] If sedulous bipartisanship is a necessary condition for the successful prosecution of modern warfare, to concede an inquiry may be an essential device of political management. Certainly immense care was taken in consulting the leader of the Labour Party (Michael Foot) and to some extent the leaders of the other parties on the fundamentals of the review: the nature and scope of the exercise (not, for example, a judicial inquiry or an investigation by a Parliamentary select committee), the choice of chairman, the composition of the membership and their terms of reference. Interestingly enough, on all four counts the Prime Minister's proposals were received with general approbation.[21]

The first and as it seems the only choice for the all-important post of chairman was Lord Franks – 'one of our most distinguished public figures', declared the Prime Minister proprietorily – a Privy Counsellor since 1949.[22] His name was originally suggested by his longstanding admirer and friend Sir Robert Armstrong, Secretary of the Cabinet, who found a receptive ear. Armstrong was promptly despatched to Franks's Oxford retreat in order to sound him out and at the same time discreetly test what has been called 'the condition of his marbles'. Any misgivings were speedily removed. The Franks family traditionally enjoys a vigorous old age. Oliver was a mere seventy-seven. His marbles, as Peter Hennessy has said, were in perfect rolling order.[23]

Franks had served inconspicuously on the shadow-play Political Honours Scrutiny Committee during Mrs Thatcher's period in office, but was not wellknown to her personally; nevertheless it is likely that in prospect she approved of him rather more strongly than he of her. In party political terms he was, in his own word, 'a neuter'.[24] Every government since Churchill's wartime coalition had availed themselves of his services in one capacity or another. Oliver Franks could be relied upon. He was sure-footed. He carried conviction and a certain moral authority – a rare quality, precociously developed. He was a man of presence and reticence, from a long line of Congregationalist ministers; a moral philosopher by training and inclination, with a magisterial cast of mind, supreme analytical skill and a brilliant expository gift. He was perhaps the wisest of a remarkable cohort of wise men thrown up by the awful opportunity of the Second World War, in his case after a meteoric rise to the top of the Ministry of Supply at the tender age of forty, with nothing but *literae humaniores* to guide him. In his diplomatic heyday as Ambassador to the United States during the momentous years 1948–52, his closest political association was with the Labour leadership of Attlee and Bevin (and their Democratic counterparts Truman and Acheson), but it was entirely in keeping that his position was unhesitatingly reaffirmed when Churchill and Eden returned to government in 1951. Throughout the period of his most active public service, from 1939 to 1962, he assiduously protected a non-party reputation, with such success that he was regularly considered for the same office by Labour and Conservative alike: most remarkably, for the Governorship of the Bank of England, a position eventually offered by the Macmillan government in 1960 and at length declined. Franks took the Liberal whip when ennobled in 1962 – he had grown up a staunch believer in the virtue of Mr Gladstone – but remained determinedly non-partisan, and continued periodically to heed the siren call of public duty until well into his eighties.

Franks was a committeeman of impressive accomplishment, and a formidable chairman. Here was someone who had chaired not only the sixteen-nation assembly which formulated the basis of the European Recovery Programme in 1947, but also the Governing Body of Queen's College, Oxford, compared to which, he has remarked, the Europeans were child's play.[25] There had been other Franks Reports: on Administrative Tribunals and Inquiries (1955–7), on Oxford University (1964–6), on Section 2 of the Official Secrets Act (1971–2) – previously, the most politically sensitive. These reports were meticulous exercises in mandarin perspicacity. They excelled in exposition and analysis, contributions to which their author attached a high value. They recommended enlightened, ameliorative reforms. They broke no moulds. In addition, Franks had sat on the committees of his only rival, Lord Radcliffe, on the Working of the Monetary System (1957–9), with Armstrong as secretary, and on Ministerial Memoirs (1975–6).

From Radcliffe he learned three lessons of general application, all rigorously observed in the Falklands Review. The first concerned the terms of reference. As he reflected on the earlier experience: 'We could have gone into all these questions [of economic policy] but, if we had produced answers to all of them, we would have been more than a committee. We would have been divinely inspired. We did not think we were divinely inspired. We therefore stuck to our terms of reference.'[26] The second concerned the outcome of the exercise as Franks saw it. In Radcliffe's report on the monetary system the eminent jurist deliberately eschewed the standard 'summary of recommendations' in favour of a strong and apposite caution:

> Our Report . . . is not to be understood as if it were a judgement passing a verdict. . . . Nor is it to be searched for the revelation of new or improved devices with prescriptions for their use, which, if applied, would produce certain and satisfactory results for the future. . . . We have noted things that might have been better done otherwise: we have said so. We have noted some cases where things should be done differently in future: we have pointed to them.[27]

The third lesson concerned the dictates of collective responsibility. 'Our view has been that, when several persons are charged with the responsibility for producing a joint report on matters that are both complex and wide-ranging, the right course is that sometimes individual opinions and preferences should be modified or abandoned for the sake of a presentation that can be put forward as the best collective view of the committee as a whole.'[28] This remained Franks's view. On the Falklands in particular, he was determined if at all possible to produce a unanimous report. In the nature of the case, that would not be easy.

Franks's colleagues were very largely selected for him. The composition of the committee was a juggling act of considerable virtuosity, much maligned. The Prime Minister suggested the idea of two Privy Counsellors from each of the main political parties, plus the former permanent secretary Sir Patrick Nairne as an 'independent' member (or a second independent member if Franks himself were so regarded). She nominated the former Conservative cabinet ministers Lord Barber and Lord Watkinson. Michael Foot nominated the former Labour cabinet ministers Lord Lever and Merlyn Rees MP.[29] All these names were mutually agreed, apparently without difficulty, though their selection faced substantial criticism, then and since. They were stigmatised as a 'comfortable, conservative and clubbable' coterie, unlikely to pursue their inquiries with the necessary zeal; greybeards of 'the governmental purple', examining the performance of their fellows and colleagues, past and present: in short members of that deplorable fraternity the Establishment.[30] And there is undeniable truth to this disparaging assessment of the committee. Privy Counsellors are honour-bound members of the Establishment: oathing is a powerful force in any tribe. Committees of inquiry necessarily draw on those who have previously lived the life: who else would sit, unpaid, unsung, and who would give attention? Reviewing the Falklands imposed an additional requirement, of especial concern to Robert Armstrong: familiarity with and sensitivity to the handling of intelligence, the last redoubt of state security.[31] From this perspective the credentials of the committee may appear, if not more impressive, then at least more appropriate. Franks had handled most things in his time, including the singular misfortune of having Philby, Burgess and Maclean all at some stage in his Washington embassy. Barber would have seen enough as Chancellor of the Exchequer, Lever as a member of the cabinet's Defence and Overseas Policy Committee, 'OD' in Whitehall parlance. The others all had some continuous experience of security matters. Watkinson had been Minister of Defence;

Rees, Northern Ireland Secretary and Home Secretary. Nairne, one of the most highly-regarded civil servants of his generation, had a long pedigree in the Admiralty, the Ministry of Defence and the Cabinet Office. Equally, his final tenure at the Department of Health and Social Security from 1975 to 1981 meant that he could in no way be associated with recent Falklands policy-making. It sometimes seemed that he had been private or permanent secretary to every important politician of the previous thirty years – among them, at the end of the 1950s, a young First Lord of the Admiralty by the name of Carrington.[32]

The disparaging assessment of the committee is more interestingly developed in what might be called the placeman theory. The theory states that 'all past prime ministers can happily sit quietly in their seats because they have a friend at court', with the corollary that civil servants had to have a friend too.[33] In this interpretation there was complete 'cover' for the entire pre-Thatcher period treated in the Report: Lever for the Labour governments of Harold Wilson (1964–70 and 1974–6); Barber for the Conservative government of Edward Heath (1970–4); Rees for Wilson's Labour successor James Callaghan (1976–9), historically, the most vulnerable of the three to detailed disinterment. Nairne seemed to be a natural choice for the civil servants' friend. Watkinson, it appeared, was friend to no one (though in office under Harold Macmillan); coincidentally, he did not take a leading part in the proceedings.[34] The prospect of an inquiry in this field ranging freely across previous administrations and rummaging through their normally inviolate papers was indeed an alarming one.[35] To the extent that the composition of the committee was designed to reassure, the placeman theory may have something to recommend it. Otherwise, like most conspiracy theories, it is appealing but overdrawn. The cardinal member of the committee, Franks himself, was left out of account. Nairne for one was miscast, his scepticism and tenacity underrated. Of the postulated connections, the most plausible was that between Rees and Callaghan, which was certainly very close. In truth Rees – mischievously described by one of his own party as 'the youngest and most militant member of the inquiry, which says a lot about its membership' – found himself in a most uncomfortable position.[36] Unlike the others, he was still an active member of a front-bench team in the House of Commons; he had to do his duty by the commit-

FIGURE 6.1 Cartoon by Peter Brooks from *The Times*, 17 January 1983. From left to right: Merlyn Rees, Lord Carrington (a witness), Lord Barber, Sir Patrick Nairne, Lord Franks, Lord Lever, Lord Watkinson. The cartoon is based on the famous painting by William Frederick Yeames, 'And When Did You Last See Your Father?'

Figure 6.2 Cartoon by Calman from *The Times*, 19 January 1983.

tee, but also by the Labour opposition. If he sought in general to defend the record of the Callaghan regime against that of the Thatcher one, he had no wish to undermine the committee by presenting a minority report. In the subsequent Commons debate, on the other hand, he felt able to vote against the report to which he had put his name, on the curious grounds that the committee had been critical enough already.[37]

Rees's colleague Lord Lever, avowedly 'nobody's man', only accepted nomination to the committee on condition that he would not have a party political axe to grind.[38] He did his bit, however, to protect Callaghan on at least one controversial issue. In November 1977, at a time of high tension in Anglo-Argentine relations, the Callaghan government famously decided to deploy a small naval force of two frigates to the area and a submarine to the immediate vicinity of the Falklands, in advance of the next round of negotiations: as the Report says, 'a force of sufficient strength, available if necessary, to convince the Argentines that military action by them would meet resistance' – that is, a measure of preventive deterrence (paragraph 65). The threat receded. In due course the force was withdrawn. Were the Argentines deterred by these means, as Callaghan himself has suggested? An important part of the answer appears to depend on the prior question of how much they knew at the time, for the deployment was made covertly. (It was revealed to the Commons by Callaghan during the crisis over South Georgia on 30 March 1982.) If Buenos Aires did not know about the deployment, the deterrent effect was at best potential and unprovable. The many unbelievers included Lord Carrington, whose officials let him into the secret on 5 March 1982 – a little late, it may be thought. Carrington asked whether the Argentines had known and was informed that they had not, whereupon 'he did not pursue the matter' (paragraph 148). For his part, Callaghan claims to have told the head of the Secret Intelligence Service, Sir Maurice Oldfield, of the decision to deploy as soon as it was taken. He hints that the information was passed on, but has provided no unequivocal statement of how and when.[39] Merlyn Rees, who was there, is certain that Callaghan arranged for the Argentines to find out about the deployment by instructing Oldfield to inform them 'in the most appropriate fashion', and that Oldfield complied; his recollection of Callaghan taking the old spymaster for a walk round the garden adds a touch of local colour to the story.[40] Harold Lever, who was also there, is equally certain that Callaghan did no such thing, or at least that the Argentines were *not* informed; his recollections was supported by a fruitless search for clues among the documents made available to the Franks committee. This negative finding may not be conclusive, but was proof enough for Lever. In the circumstances he was anxious to minimise any embarrassment to Callaghan.[41] Thus the deadpan first draft of the Report, 'We have had no evidence that the Argentine Government were informed of this deployment,' gave place to a final version in which the evasion '*became aware*' was deftly substituted for 'were informed'.[42]

According to its many detractors, such a committee could be expected to produce only one kind of report. 'In the Franks Report,' wrote Correlli Barnett, 'the British Establishment has sat in judgement on the British Establishment and found it not guilty.' On this view the Report, like the committee, was fatally compromised. It was 'a classic Establishment job', in the words of its most trenchant critic, Hugo Young.

> Lord Franks' admirers saw the report as his last flawless exercise in mandarin ambiguity. But this was not its political impact, and nor, one must surmise, was it intended to be.
>
> What the Franks Report was really addressing was another imperative familiar to mandarins, that of political reality. The fact was that the war had been won, and nothing could be allowed to interfere with this great event. Had the war been lost, the same set of facts would have been produced as a devastating proof of negligence. But, as was . . . later conceded from inside the committee, Lord Franks' strategic objective was to ensure that Mrs Thatcher's reputation should not be damaged. He could see no possible need, in the circumstances, for any other course of action.[43]

This is a grave indictment. It rests on the apparent exoneration offered in the conclusion of the Report, the notorious paragraph 339, or more precisely in the latter part of its last sentence, which by a quirk of fate appeared in splendid isolation on the final page. The offending passage runs: 'we conclude that we would not be justified in attaching any criticism or blame to the present Government for the Argentine Junta's decision to commit its act of unprovoked aggression in the invasion of the Falkland Islands on 2 April 1982'. Anticipating Hugo Young, Simon Jenkins has charged that 'the only basis for this exoneration – admitted privately by some Franks committee members – was an understandable desire not to reopen political wounds at a time of rejoicing'.[44]

The immediate impression of a final verdict delivered as a ringing declaration of faith in the victress received neat visual reinforcement from the suggestive lay-out of the final page of the Report. The headlines were set accordingly. 'Franks clears Maggie' was the reflex reaction, a reaction conditioned by the artful news management of the Report's rush release: a tightly restricted distribution of advance copies, a prepared list of the numbers of 'key paragraphs', a spate of pre-publication leaks studiously designed to discount criticism of the Prime Minister herself and implant the idea of a long period of cross-party war guilt.[45] As to the source of the leaks, Franks was reminded of a remark by Harold Macmillan: 'Political casks always leak at the top'.[46] The spin doctors had done their work; the ground had been prepared. The most celebrated variant of the 'Franks clears Maggie' approach was James Callaghan's well-made jest that 'for 338 paragraphs the Franks Report painted a splendid picture, delineating the light and shade. The glowing colours came out. When Franks got to paragraph 339, he got fed up with the canvas that he was painting and chucked a bucket of whitewash over it'.[47] The cartoonists had a field day.

Without subscribing to mandarin ambiguity, it is perfectly possible to see in the Report's conclusion something more than a political slogan. The paragraph should be taken whole:

> Against this background [a recapitulation of the constraints on policy-making] we have pointed out in this Chapter where different decisions might have been taken, where fuller consideration of alternative courses of action might, in our opinion, have been advantageous, and where the machinery of Government could have been better used. But, if the British Government had acted differently in the ways we have indicated, it is impossible to judge what the impact on the Argentine Government or the implications for the course of events might have been. There is no reasonable basis for any suggestion – which would be purely hypothetical – that the

FIGURE 6.3 Cartoon by Lurie from *The Times*, 19 January 1983. 'It was a thorough clean-up inside and out.'

FIGURE 6.4 Cartoon by Peter Brooks from *The Listener*, 27 January 1983.

> invasion would have been prevented if the Government had acted in the ways indicated in our Report. Taking account of these considerations, and of all the evidence we have received, we conclude that we would not be justified in attaching any criticism or blame to the present Government for the Argentine Junta's decision to commit its act of unprovoked aggression in the invasion of the Falkland Islands on 2 April 1982.

The first thing to be said about this paragraph is that it was drafted collectively in committee: it was a composite effort, strenuously achieved. This was highly unusual. The bulk of the drafting was usually done by the secretary, Anthony Rawsthorne, selected partly for this purpose; most of the rest by the fabled Patrick Nairne. Franks contributed a detailed scrutiny of the text but habitually drafted little. The final paragraph must have seemed too important to be left to the secretary. There may have been varying motives for accepting what ultimately emerged. Political loyalties naturally tended to reassert themselves. Different committee members attached special significance to different parts of the whole. For Rees, the first sentence was the key: it reiterated the Report's critique of the government in power. For Nairne, as we shall see, the second and third sentences were crucial. For Barber and Watkinson, the last sentence accommodated their own reluctance to weaken Margaret Thatcher.[48] It was for the chairman, who was wont to exercise a tight rein on the proceedings, to ensure that collective responsibility ultimately prevailed.

How then did the committee reach this conclusion, and what exactly did it mean? *Pace* Hugo Young, Franks himself 'had no motive or desire to move one way or the other', and no grand design. 'I hadn't a clue where we would come out.'[49] Mrs Thatcher's reputation was not his prime concern. Of course he was alive to the political explosiveness of the inquiry. He questioned Armstrong closely at the outset on the kind of job that the committee was being asked to do. The advice he received was to provide a definitive statement which would take the issue out of domestic party politics and 'turn it into political history'.[50] These were fundamental considerations from an influential source, but they gave no more than general guidance. Franks was neither predisposed nor preordained to acquit.

The committee spent their first six weeks combing the unprecedented mass of government documents made available to them. Unlike the Dardanelles Commission, each member went through everything, individually. Apart from the direct benefits of this procedure, Franks wanted to impose a pause in order to defuse any partisan feeling in the immediate aftermath of the war. He also wanted to be quite certain that there had been full disclosure by the departments concerned, going so far as to check the serial numbers of telegrams and request written assurances from the respective permanent secretaries.[51] Only after an intensive period of reading-in and discussion did the committee begin taking oral evidence. Franks was a past master at this, and a large number of prominent witnesses had the benefit of a searching cross-examination, conducted with matchless punctilio. They included the Prime Minister herself, exhaustively prepared, who performed keenly to scholarship standard.[52] The richest evidence, however, came from the documents. In forming their assessment the committee followed three self-imposed injunctions: to avoid the exercise of hindsight, to determine what was reasonable in the light of the circumstances prevailing at the time, and to allow for the fact that those involved were not dealing exclusively with the Falklands (paragraphs 13 and 14). This approach, fundamental to the outcome of the Report, has been found too lenient or plain misguided. 'If you examine people's actions "according to their own lights" you are bound to exonerate them.'[53] Yet it is difficult to see how else the inquiry could have proceeded. 'Reasonable men', a phrase with a Shakespearian echo, may have been a construct of the supremely reasonable Lord Franks, but the test of *un*reasonable judgement or behaviour remained a sensible criterion for appraising the discharge of the responsibilities of government. More arguable were the two questions formulated by the committee in order to make that appraisal. Could the government have foreseen the invasion on 2 April? And could the government have prevented that invasion (paragraph 260)? The first was answered categorically in the negative. 'The invasion of the Falkland Islands on 2 April could not have been foreseen' (paragraph 266). There is a misconception that the second was answered similarly. In fact it was not. 'No simple answer' was available. The Report merely indicated that 'it is impossible to judge' whether an invasion on that date could have been prevented (paragraphs 336 and 339).[54]

On the basis of this approach and these questions and answers the Franks Report concluded as it did. The questions it posed – questions of the committee's own devising – were highly specific. The answers it gave were carefully circumscribed. Both related to one particular date. Contrary to reflex reaction, the Report's 'exoneration' of the Thatcher government was, by deliberate intent, materially incomplete. As always Franks stuck strictly to his terms of reference. Put crudely, the Report blamed General Galtieri rather than Mrs Thatcher (or Lord Carrington) for what happened on 2 April 1982. No explicit conclusion was drawn about what had happened earlier or what might have happened later. As Sir Patrick Nairne has explained,

> we were not asked to review the discharge of the government's responsibilities in relation to the *general deterrence of Argentina* from invasion. Of course, the report says a lot about deterrent action, and we criticised the government for some inadequacies and failings; but here the second and third sentences of paragraph 339 are crucial – and we might perhaps have given them greater emphasis. If the strategic deployment by *successive* governments of our naval forces had given the South Atlantic a higher priority, and if in 1981–82 we had not been trying to persuade Argentina to 'cool it', Argentina *might* have been deterred from invasion; but, in the actual circumstances, the committee could not conclude that the Thatcher government should be blamed for Galtieri's dash for the Malvinas.[55]

This explanation is cogent but evidently frustrating. There is a lingering feeling that by fixing his attention on 2 April Franks succumbed to an over-literal interpretation of his terms of reference. But the principal cause of the frustration is a morbid fascination with the concluding paragraph.[56] That unhappy composite was the terminus and not the focus of the Falklands Review. 'It is essential that our Report should be read as a whole', the committee protested (paragraph 336). Their protestation was ignored. Paragraph 339 became the cynosure of the Report: truly a case of the tail wagging the dog.

The contributory causes are more speculative. Perhaps the very terms of reference were flawed – deceptively open, yet framed knowingly or unknowingly to dam the flow of the inquiry. The suggestion is as intriguing as it is imponderable. Perhaps the committee's questions were deficient. Numerous alternatives have been put forward; often they were subsumed in the rest of the Report.[57] Certainly the answers repay some study. The invasion was impossible to foresee, it was argued, because of the close timing of the decision. Prevention was difficult and uncertain because at any given moment the conceivable measures might be too provocative or too late. The committee estimated that the final decision to invade was not taken until 31 March 1982 at the earliest, though they noted correctly that an Argentine task force had put to sea on 28 March (paragraph 263). We now know that the decision was taken rather sooner, on 26 March. Firm Argentine evidence on this point was not then available. More importantly, the junta had been hatching schemes since January 1982 for such an operation later in the year; they approved an outline plan on 9 March and called for an immediate consolidated version on 23 March.[58] Concentrating on the British side, the committee monitored the quickening tempo of events in February/March 1982 and criticised the Foreign Office sharply for failing adequately to respond (paragraph 302). Franks himself thought that there was a tendency for officials to be imprisoned by the continuum of their past, and therefore to experience difficulty in reading the multiple signs of a gathering storm.[59]

Yet although the Report properly identified the South Georgia crisis as pivotal, its analysis of British moves in late March 1982 is often wanting. This relates particularly to their unintended consequences, for it is the *interactive* nature of the crisis that impresses. South Georgia was in that respect a microcosm of the wider conflict. The decision to despatch the venerable *HMS Endurance* there from Port Stanley on 21 March and the so-called 'Carrington ultimatum' of 23 March, for instance, may not have been 'war-starters' (Admiral Anaya's telling phrase) but were almost certainly crisis-stimulators, inducing the junta to accelerate their own timetable of operations.[60] The subsequent decision to despatch a nuclear-powered submarine to the Falklands on 29 March, to be on station some two weeks later – a decision leaked almost immediately by the *Daily Express* – can only have strengthened their resolution to seize the day.[61] A series of 'final warnings' from London, woefully tepid, scarcely gave them pause. On 23 March a junior minister stated in the Commons that it was the duty of any government 'to defend and support the Islanders to the best of their ability'. On 25 March the Argentine Foreign Minister was informed that Britain was 'committed to the defence of its sovereignty in South Georgia as elsewhere'. On 31 March President Reagan was asked to make clear to the Argentine government that Britain 'could not acquiesce' in action against the islands, an instruction he comfortably exceeded by speaking in terms of a *casus belli* when eventually he got through to Galtieri the following day – too late.[62]

For the Falklands, it was usually too late. 'Has it ever occurred to you that nothing is ever done until everyone is convinced that it ought to be done, and has been convinced for so long that it is now time to do something else?'[63] The story that Franks elected to tell was a modern tale of knight-errantry. Great deeds were done and great opportunities missed. There was a certain incoherence to the action. The plot drifted, helplessly. A pervasive air of artificiality was punctured by events of pure magic realism. Read from the beginning, the Franks Report is a marvellous chronicle of unripe time.

'Timing' dominates the Report. For successive British governments of varying political hues time was lost and gained, bought and sold, stolen and retrieved, right and wrong, sometimes needed and sometimes not.[64] Time might run on or run out, but in the years covered by the Franks Report it was never, ever, ripe. Most fundamentally, this applied to the collective formulation and concertation of ministerial views, specifically in OD, the mythic cabinet Defence and Overseas Policy Committee, whose signal failure even to discuss the Falklands from January 1981 until the day before the invasion is the most remarkable feature of 'the nature of government' exposed with such startling clarity by the Report. If the first responsibility for that failure was the Foreign Secretary's, the final responsibility was clearly the Prime Minister's. It is scarcely necessary to read between the lines of the Report – a technique recommended somewhat disingenuously by several commentators – to trace the process of decision-making to its source. Significantly, the Report also disclosed that during the same period there was no reference to the Falklands in full cabinet until Carrington related the events in South Georgia on 25 March 1982. For the attentive reader it is all there, 'plain as a pikestaff'.[65]

Why was there no meeting of OD? Because 'the time was never judged to be ripe' (paragraph 291). This was the luckless Lord Carrington's judgment. Carrington's permanent secretary, Sir Michael Palliser, who was not the only official to favour such a meeting – in September 1981 for example – has since ventured an explanation of his master's thinking:

> Lord Carrington just felt that the *timing* was wrong. That reflected not in the least hostility to him *in the cabinet* – I really don't think that was there at all – but more a feeling that the only time when one would really get a British cabinet to *focus* on the issue was at a time when it was manifestly going to cause real difficulty if they did not.[66]

Carrington had known what ought to be done since 1979, when he first recommended substantive negotiation with Argentina on sovereignty, with leaseback as the end in view. The Prime Minister decreed that Rhodesia must be settled before there could be any discussion of the Falklands.[67] Carrington delivered a Rhodesian settlement, in the process alienating not a few in his own party, a major source of the hostility mentioned by Palliser. By 1981 he had not changed his mind about what ought to be done, but he also knew that in order to do it he would need to carry his principal cabinet colleagues with him, and that public (and islander) opinion would have to be 'guided' or 'educated' to accept it – to say nothing of the difficulty of the diplomatic negotiations themselves. He realised that he had no hope of meeting even the first of these requirements.[68] There was in existence a profound suspicion of the Foreign Office, to which the Prime Minister was by no means immune, well characterised by Carrington himself in retrospect as a belief 'that there is something disreputable or even treacherous in trying to seek agreement with foreign governments'. Holders of this belief 'carry chauvinism and insularity to such a degree that one almost feels they disapprove of anyone in the Foreign Office talking to a foreigner. Negotiation, it seems, is feebleness and unpatriotic. But the alternative to negotiation is confrontation.'[69] Confrontation was the province of the Secretary of State for Defence, John Nott, Carrington's chief interlocutor on the issue. As Sir Anthony Williams, then Ambassador to Argentina, later remarked, the problem of defending the Falklands – more realistically, perhaps, deterring the Argentines – was not a Foreign Office problem. 'It was a Ministry of Defence problem, and particularly it was a Ministry of Defence *budget* problem. We [the diplomats] were asked to operate this Argentina difficulty within certain parameters. One of those parameters was that, actually, we had no money to spend. Indeed, such money as we were spending was being cut down.'[70] Public expenditure cuts were the order of the day. In the midst of a stringent Defence Review, locked in mortal combat with the Chiefs of Staff, John Nott proved a singularly unaccommodating colleague.

Some six months before the invasion, therefore, Carrington decided definitively to perpetuate an old British cross-party tradition. He thought he had a little time. Hopefully, he opted to play for more. He would neither break off negotiations nor promote leaseback. For the Foreign Secretary it was a doomed gamble on a dwindling asset. As Michael Charlton has written, 'these were the months which the locusts ate'. From remote Buenos Aires Anthony Williams lodged his celebrated protest over a decision 'to have no strategy at all beyond a general Micawberism'.[71] The object of the game slipped elusively away. On 5 March 1982 Carrington declined to pursue the matter of a naval deployment. 'The last moment at which . . . it would have been possible to sail a deterrent force to be in place' and ready, according to Franks, was now past (paragraph 300). 'The psychological moment' for which Carrington was waiting to face his colleagues with an incontrovertible policy choice never came.[72] On 2 April 1982 time abruptly ran out. For diplomacy and defence alike, the moral of the Franks Report is an ancient one. 'Time, by the way, is like the medlar; it has a trick of going rotten before it is ripe.'[73]

NOTES

I am grateful to Peter Hennessy, Dan Keohane and Sir Patrick Nairne for their comments on earlier drafts of this chapter.

1. F.M. Cornford, *Microcosmographia Academica*, 3rd ed. (Cambridge: Bowes and Bowes, 1938) p. 32.

2. Carrington interviewed by Robert Kee on BBC TV 'Panorama' programme, 5 April 1982, printed in *The Times*, 18 January 1983; *Reflect on Things Past* (London: Fontana, 1989) p. 370. The *Economist* wrote of 'the hounding from office' of the ministers in question. 'The misjudgements that led to the Falklands war', 19 June 1982.
3. *Official Report* (House of Commons unless otherwise indicated), 8 April 1982, col. 416.
4. Ibid., 6 July 1982, cols. 51–2.
5. Cmnd. 8787, *Falkland Islands Review* (London: HMSO, 1983) (hereafter Franks Report); interview with Anthony Rawsthorne, 15 March 1990. Edited versions of the Report's concluding chapter (paras. 260–339) appear in the *Economist*, 22 January 1983, and in Max Hastings and Simon Jenkins, *The Battle for the Falklands* (London: Pan, 1983) pp. 417–33.
6. W.J.M. Mackenzie, 'The Plowden Report: a translation' (1983), reprinted in Richard Rose (ed.), *Policy Making in Britain* (London: Macmillan, 1969) pp. 273–82. See para. 3. The original was Cmnd. 1432, *Control of Public Expenditure* (London: HMSO, 1961).
7. *Official Report*, 25 January 1983, col. 831.
8. Cf. Nigel Clive, 'The management of intelligence', *Government and Opposition* 22 (1987) pp. 93–100; William Wallace, 'How frank was Franks?', *International Affairs* 59 (1983) pp. 453–8.
9. *Official Report*, 8 July 1982, col. 469.
10. On the lack of precedent see, e.g., Lords Thomas and Hatch, *Official Report*, House of Lords, 25 January 1983, cols. 208 and 212. On the Dardanelles Commission see Martin Gilbert, *Winston S. Churchill*, vol. III and companion vol., part 2 (London: Heinemann, 1971 and 1972) pp. 789ff. and 1560ff. Cf. Cmnd. 8490, *The Final Report of the Dardanelles Commission* (London: HMSO, 1917).
11. Churchill to the Commission, quoted in Gilbert, *Churchill*, IV, p. 9.
12. For their personal views on various aspects of the committee's deliberations I am especially grateful to Lord Franks, Lord Lever, Sir Patrick Nairne, Anthony Rawsthorne, and Merlyn Rees. Lord Watkinson makes brief reference to its work in his memoirs, *Turning Points* (Salisbury: Michael Russell, 1986) pp. 214–18.
13. Churchill in the Commons, quoted in Gilbert, *Churchill*, IV, p. 11.
14. Franks Report, Annex A, Assertion 2.
15. On the prehistory, Peter Beck, *The Falkland Islands as an International Problem* (London: Routledge, 1988) pp. 88ff.; on the urgency, *Official Report*, 8 July 1982, col. 473; Franks Report, para. 7; interviews with Lords Franks and Armstrong, 16 February and 8 March 1990.
16. 'The Franks Report', editorial, *The Times*, 19 January 1983.
17. They were Sir Frederick Cawley, a Liberal MP; James Clyde, a Liberal Unionist MP; Captain Stephen Gwynn, a journalist and Nationalist MP; Admiral Sir William May; Field Marshal Lord Nicholson; Sir William Pickford, a Lord Justice of Appeal; and Walter Roch, a Liberal MP.
18. Franks interview, 16 February 1990. The substance of the Privy Counsellor's oath is given in Halsbury's *Laws of England*, 4th ed. (London: Butterworths, 1974) vol. VIII, p. 708.

19. *Official Report*, 8 July 1982, col. 484. Cf. James Callaghan and a delphic Edward Heath, ibid., cols 479 and 493.
20. 'Walking on water', *Economist*, 22 January 1983. On being 'forced', see George Foulkes, *Official Report*, 26 January 1983, col. 971.
21. *Official Report*, 8 July 1982, cols. 469–72, and 26 January 1983, cols. 995–6.
22. Ibid., 26 January 1983, col. 995.
23. Peter Hennessy, 'The lord who sits in judgement', *The Times*, 17 January 1983; Armstrong interview, 8 March 1990.
24. Franks interview, 16 February 1990. The following paragraphs draw on my study of Franks, *Founding Father* (Oxford: Oxford University Press, forthcoming). See also Peter Hennessy, *Whitehall* (London: Secker and Warburg, 1989) pp. 568–74.
25. Franks interview, 15 September 1988. On Franks in committee see Alex Danchev, 'Taking the Pledge: Oliver Franks and the negotiation of the North Atlantic Treaty', *Diplomatic History* 15 (1991).
26. Oliver Franks, *Some Reflections on Monetary Policy* (London: Asia Publishing House, 1960) p. 3.
27. Cmnd. 827, *Report of the Committee on the Working of the Monetary System* (London: HMSO, 1959) para. 979.
28. Ibid., para. 15.
29. *Official Report*, 26 January 1983, cols. 995–6. Rees had also been a member of the committee on Section 2 of the Official Secrets Act.
30. Dick Douglas, *Official Report*, 8 July 1982, col. 472; Hugo Young, 'An establishment job', *Sunday Times*, 23 January 1983. For Franks's view of the Establishment see Hennessy, *Whitehall*, p. 545.
31. *Official Report*, 8 July 1982, col. 469; Armstrong interview, 8 March 1990.
32. Interview with Sir Patrick Nairne, 16 February 1990. Cf. Carrington, p. 147.
33. Alexander Lyon, *Official Report*, 8 July 1982, col. 498. Cf. 'The Franks Report', *The Times*, 19 January 1983.
34. Private information. Macmillan, unimpressed, despatched Watkinson in the 'night of the long knives' in July 1962. See Alistair Horne, *Harold Macmillan*, vol. II (New York: Viking, 1989) pp. 341, 345.
35. See Edward Heath, *Official Report*, 8 July 1982, cols. 495–6.
36. George Foulkes, ibid., col. 491.
37. Interview with Merlyn Rees, 5 February 1990. The connection with Callaghan is clearly documented in Rees's entry in Andrew Roth's *Parliamentary Profiles* (London: Parliamentary Profiles Service, 1984) pp. 661–2. See also James Callaghan, *Time and Chance* (London: Collins, 1987) pp. 234–5, 387.
38. Interview with Lord Lever, 5 February 1990.
39. Lord Callaghan, letter to the author, 7 June 1990. Cf. Callaghan, *Official Report*, 26 January 1983, cols. 947–8; *Time and Chance*, p. 375; interviewed by Michael Charlton in *The Little Platoon* (Oxford: Blackwell, 1989) p. 65.
40. Rees interview, 5 February 1990.
41. Lever interview, 5 February 1990.
42. Franks Report, para. 66 and Annex A, Assertion 5. For other firsthand

testimony see Benn diary, 15 November 1977, printed in Tony Ben, *Conflicts of Interest* (London: Hutchinson, 1990) pp. 248–9; Denis Healey, *The Time of my Life* (London: Michael Joseph, 1989) p. 494; David Owen, *Official Report*, 25 January 1983, cols. 817–18, and 'How Franks missed the real point', *Observer*, 23 January 1983; and Ted Rowlands, 'Five years before the invasion', *The Times*, 15 January 1983. The Argentine evidence is equally contradictory. For further analysis see Alex Danchev, 'Life and death in the South Atlantic', *Review of International Studies* 17 (1991).

43. Correlli Barnett, letter to *The Times*, 26 January 1983; Hugo Young, *One of Us* (London: Macmillan, 1989) p. 284, foreshadowed in 'An establishment job', *Sunday Times*, 23 January 1983. Cf. his very similar indictment of the Calcutt Report on Privacy (1990): 'The great, the good and the disastrous', *Guardian*, 26 June 1990. See also Anthony Barnett, 'To be absolutely Franks', *New Statesman*, 21 January 1983; Simon Jenkins, 'The pardoner's tale', *Spectator*, 2 April 1983.
44. Simon Jenkins, 'Britain's Pearl Harbour', *Sunday Times*, 22 March 1987.
45. Philip Webster, 'Poor intelligence blamed by Falklands report', *The Times*, 17 January 1983; Michael White, 'Anger at whispering campaign on Franks', *Guardian*, 17 January 1983; *Official Report*, 18 January 1983, cols. 171–6.
46. Franks interview, 16 February 1990.
47. *Official Report*, 26 January 1983, col. 945.
48. Rees, Nairne and Rawsthorne interviews, 5 February, 16 February and 15 March 1990.
49. Franks interview, 16 February 1990.
50. Armstrong interview, 8 March 1990.
51. Franks Report, paras. 4–7; Franks and Rawsthorne interviews, 16 February and 15 March 1990.
52. Franks interview, 16 February 1990. Confirmed by other members of the committee.
53. Barnett to *The Times*, 26 January 1983. Cf. G.M. Dillon, *The Falklands, Politics and War* (London: Macmillan, 1989) pp. 53–4; Lord Wilson of Langside, *Official Report*, House of Lords, 25 January 1983, cols. 217–18.
54. Cf. 'Walking on water', *Economist*, 22 January 1983. Lever and Rees, especially, were keen to argue that the question of prevention was an unanswerable one. 'If I had agreed to my wife having a blue hat, not a red one, the marriage would have gone well.' Lever interview, 5 February 1990.
55. Sir Patrick Nairne, letter to the author, 3 March 1990. The Report was silent on Carrington's resignation.
56. Virtually the only commentator to wrestle free of para. 339 was Peter Jenkins (an opponent of the war), whose immediate verdict on the Report was exceptionally favourable. 'An acquittal but a grave indictment', *Guardian*, 19 January 1983.
57. See the criticisms of Denis Healey, *Official Report*, 26 January 1983, col. 931; and the suggestions of Lords Hatch and Beloff, *Official Report*, House of Lords, 25 January 1983, cols. 212 and 222.
58. Admiral Carlos Büsser interviewed by Charlton, pp. 111–15; O. Cardoso

et al., *Falklands – the secret plot* (East Molesey: Preston Editions, 1987) pp. 72–6; Lawrence Freedman and Virginia Gamba-Stonehouse, *Signals of War* (London: Faber, 1990) pp. 52ff.; Martin Middlebrook, *The Fight for the 'Malvinas'* (London: Viking, 1989) pp. 1–25.

59. Franks interview, 16 February 1990.
60. Middlebrook, p. 13; Admiral Harry Train USN (who debriefed Anaya for a classified American study of the war) in Charlton, p. 116.
61. Cf. Anaya's reaction to (unidentified) reports of an earlier deployment, related by Train in Charlton, pp. 116–18.
62. Franks Report, para. 333; Healey, *Official Report*, 26 January 1983, col. 937; Hastings and Jenkins, p. 380. The transcript of the Reagan-Galtieri conversation is printed in Cardoso, pp. 83–6.
63. Cornford, p. 4.
64. For classic statements of the need to 'play for time' (in January 1977) and to 'buy time' (in February 1982) see Hugh Carless and Richard Luce interviewed by Charlton, pp. 60–1 and 182–3.
65. The phrase is Franks's. Interview, 16 February 1990.
66. Palliser interviewed by Charlton, p. 138.
67. Franks Report, paras. 73–6.
68. Carrington, pp. 348–64. Sir Nicholas Henderson's shrewd commentary on the Franks Report noted that it did not confront the issue of 'education'. *Sunday Times*, 23 January 1983.
69. *Official Report*, House of Lords, 25 January 1983, col.160.
70. Williams interviewed by Charlton, p. 127.
71. Charlton, p. 125; Franks Report, para. 104.
72. Palliser quoted in *Official Report*, 26 January 1983, col. 935.
73. Cornford, p. 32. The medlar is a small fruit tree.

7 Intelligence Warning of the Occupation of the Falklands: Some Organisational Issues

Michael Herman

All modern states rely to some extent on their intelligence organisations, but the service they get from them is curiously uneven. Sometimes they are war-winners, as was the case in World War II: the British official historian has claimed that the Western Allies' codebreaking successes shortened the war by three or four years.[1] Yet on other occasions the possession of an apparently efficient intelligence machine seems no guarantee against misperception. This is specially true of warning of surprise attack. There have been a whole series of failures in warning of surprise military attacks, ranging from the North Korean attack on South Korea in 1950 through the Egyptian and Syrian attacks on Israel in 1973 to the Iraqi attack on Kuwait in 1990.[2] Despite all the paraphernalia of modern warfare and the intelligence technology available to detect it, warning is evidently still a chancy business.

An important question for national defence and international stability is, therefore, the weight to be attached to intelligence in this warning role. Are its failures inherent in the nature of international relationships, or do they result from a normal and to some extent remediable quota of human frailty? If warning performance can be improved, what are the recipes? The answers may be relevant to other, wider ways in which governments use their intelligence services. If intelligence cannot be relied upon for warning of military attack, perhaps it is equally fallible elsewhere. The structure built up to reduce national risks may actually increase them.

Commentators have grappled in various ways with this issue of reliability. At one extreme there is the rubbishing school, depicting intelligence as an example of self-serving bureaucratic growth:

> So rather like drunken sailors propping each other up, the intelligence agencies of East and West need each other as an excuse to justify this aspect of their activities [espionage].... In post-war years Britain's security service has seldom caught a spy or traitor on its own, while the IRA is no nearer defeat after 20 years of intensive security operations.[3]

At the other, there are the apologists and enthusiasts who, in the spirit of the motto on the CIA building, 'And ye shall know the truth and the truth shall make you free,' see intelligence as the indispensible key to wise policy:

> We believe that to be successful in its foreign policy, a nation – whether democratic, autocratic, or totalitarian – needs accurate political, economic, and military intelligence regarding its rivals, its competitors, its enemies, and – yes – even its friends.[4]

Somewhere between these two extremes are the academics who have produced rather less black-and-white interpretations. Strategic surprise usually succeeds, as does deception: 'Surprise is almost always unavoidable – and will continue to be so in the foreseeable future – despite all efforts to the contrary.'[5] Deception also has the odds stacked in its favour; in sixty-eight cases of it between 1914 and 1968 there is said to have been a 91 per cent success rate.[6] There is not much that can be done to overcome cognitive biases: 'successful detection of military and diplomatic surprise is less likely to be explained by the skill of the intelligence service in piecing together arcane bits of information than by the degree to which the service's predisposition and expectation fit the actions that the other is planning to undertake'.[7] Modern largescale intelligence organisation introduces its own pathologies into the warning process:

> collected information must pass through numerous potential bottlenecks. It must be screened at low levels to raise an initial suspicion; it must be transmitted to higher levels of the intelligence bureaucracy to be compared with data from other

sources; it has then to be passed to policymakers who must judge whether the evidence warrants action.[8]

Even if intelligence perceives the truth the policy-makers may disregard it, or may pick the interpretations that suit them – 'Interested policy-makers quickly learn that intelligence can be used the way a drunk uses a lamp post – for support rather than illumination.'[9] Or intelligence may not be consulted at all, as over the Israeli decision to invade the Lebanon in 1982.[10]

The general conclusion from this academic critique is that intelligence has a role, but more as an intellectual critic of policy preconceptions than as a purveyor of 'truth':

> Intelligence may not be able to find the truth; even less may it be able to persuade others that it has found it. But 'keeping the players honest,' not permitting disreputable arguments to thrive, pointing out where positions are internally contradictory or rest on tortured readings or the evidence would not be a minor feat. While it would not save the country from all folly, it would provide more assistance than we get from most instruments of policy.[11]

The moral is *not* to gear policy too closely to it; policies that depend on completely accurate intelligence 'are likely to yield disaster'.[12]

These conclusions draw heavily on the record of misperception and warning failures, and the events leading up to the Falklands invasion form part of the canon. *Au fond,* the war may have been a rational conflict fought for identifiable reasons on both sides, each with a reasonable chance of succeeding; but the route to it was via intelligence and policy fallibilities. The junta assessed that the British would not fight to repossess the islands: the British assumed over a long period that if there was Argentine military action it would be preceded by graduated diplomatic pressure and rising tension. On the British side there were also what can be seen (with hindsight) as failures in the warning machinery. The diplomatic evidence of hardening Argentine attitudes was not given the weight that it deserved; other information may not have been sought with sufficient priority; the central machinery was curiously slow off the mark during the events leading up to the invasion. The American intelligence community seems to have been similarly unprepared.

We probably still need more information before drawing conclusions about the Argentine and American performance. But on the British side the Franks Report (discussed elsewhere in this book by Alex Danchev) makes a major contribution to ideas about intelligence warning.[13] These ideas are drawn on by historians to understand the historical events, and one might also hope that they are drawn on by practitioners for the evolution of practical doctrine. Debate about the intelligence lessons of the Franks Report is therefore not unimportant. This commentary limits itself to some brief remarks on the significance of British organisational factors, and on the general relationship between organisation and intelligence failure.

Some preliminary points can be made in passing. In the first place, longer-term perceptions are inseparable from current surveillance, and the quality of the one determines the quality of the other. The failures over the Falklands were in not focussing on the problem early enough in 1982, because the received long-term view was that the Argentine government would apply political and economic pressure before attempting a military solution. Secondly, Argentina was (and indeed remained) a relatively small-scale intelligence target, very different from those of the cold war. (Indeed one of the issues raised by the Falklands experience – even more relevant to the 1990s than the 1980s – is how intelligence organisations can best keep an effective eye on small targets well below the top of any rational priority ladder.) Thirdly, the uniqueness of the Franks Report as an account of the British machinery should not lead to its use as a template for all occasions. The Argentine invasion of the Falkland Islands in 1982 was essentially a *coup de main*, rather than a surprise military attack as in 1950 or 1973 or a well-prepared military occupation like the Soviet invasion of Czechoslovakia in 1968. The inadequate weight given by the British assessment machine to the diplomatic evidence was *not* typical, despite the attention that has been subsequently drawn to it.[14] A far more common cause of failure has been to overemphasise reassuring political indicators at the expense of worrying (but usually ambiguous) military indicators.[15]

The problems of warning go rather deeper, and involve the pattern of intelligence organisation, the structure of the intelligence community, and the policies for its recruitment and manning. There are at least five dilemmas, to which each national system has its characteristic responses. Thus, *first*, intelligence analysts have to be sufficiently close to their readers to give them credibility and a sense of relevance and timing, yet sufficiently distanced from them to maintain objectivity. *Second*, analysis can either be organised around a central agency of some kind, or based on departmental organisations brought together through some federal interdepartmental structure. *Third*, intelligence assessments can represent the views of intelligence specialists, uninfluenced by policy considerations, or can be the means of welding together *all* sources of governmental knowledge, including those of foreign ministries and other 'operating' departments. *Fourth*, control of the intelligence community can be on a tight or loose rein, either with tight central control or (more commonly) considerable departmental autonomy. *Fifth*, staffing patterns can emphasise either the role of 'professional' intelligence officers, or the use of people on secondment who bring the professional skills of diplomats, economists or service officers to the business of intelligence assessment.

This is not the place to explore these choices in detail, but some features of the British system can be indicated by its place on these axes. Warning depends upon getting all the relevant evidence together in one place, and evaluating it with the aid of a good institutional memory and the relevant long-term preconceptions. Britain has no single intelligence agency like the CIA responsible for studying foreign countries as a whole, either for warning or for any other purposes. The Defence Intelligence Staff (DIS) is the nearest thing to a comprehensive analysis agency, but its focus is primarily on military matters and those things which contribute to foreign military power.

To bring political, military and other strands of evidence together into all-round assessment the system therefore depends heavily on the work of the Assessments Staff in the Cabinet Office and the committee work of the Joint Intelligence Committee (JIC) and its subordinate groups. The work of the Assessments Staff is fused with that of JIC member departments; the results are subject to discussion by the intelligence community and are presented as consensus judgments. The model is therefore a 'modified federal' one, depending on the high profile central staff to provide coordination, leadership, and independent scrutiny of departmental positions.[16] However, the Assessments Staff is quite small, and has few facilities for research or institutional memory. Yet no intelligence judgment can be better than the research and the data base available on the subject, and for these the Assessments Staff depends on support from the rest of the intelligence community.

This departmental effort is uneven. On things associated with military power the DIS does the basic analysis and research, but is variable in its quality – reflecting problems over the incidence of short-term military postings and the recruitment and retention of high-quality professionals. On political intelligence, by contrast, there are plenty of brains but not much nitty-gritty research. The Foreign and Commonwealth Office (unlike the US State Department) does not have a department for intelligence analysis, but relies on its operational departments to assess information as it reaches them and to participate in the work of the central assessment machinery. (It does indeed possess its relatively small Research Department, but this has tended to be distanced from current intelligence production.) Embassy reporting and FCO analysis is of good quality, but is oriented towards the practical and immediate side of international relations, rather than scanning the horizon for distant threats.

The JIC system relies on the central staff and the committees to compensate for such unevenesses among the participating departments. Both the staff and the system run on an informal, consensus-seeking operational style; the club-like metaphor of the 'intelligence community' is significant in itself. There is little of the formal management apparatus of directives, targets, management reviews, appraisals and the like. The system runs by flexibility, initiative and seat-of-the-pants judgments. The assumptions are that departments will not fight their corners beyond a certain point, and that good people can be trusted to act sensibly without close supervision.

This accentuates the other unevenness: the variable influence of professional analysts compared with those on secondment. On the whole, Britain since World War II has invested in professionalism at the collection end of intelligence, rather than in mainstream analysis. In pay and status the civilian analysts of the DIS are the underclass of the British intelligence community. Many of the organisation's posts – including most of the senior ones – are filled by service officers; and since the services themselves have failed to develop intelligence career structures most of them are doing their first and only intelligence tours. Within the Assessments Staff, most of its members are also on secondment, mainly from the FCO and the services. The professional diplomats are trained analysts of foreign countries and have no difficulty with most types of intelligence assessment, but they lack the background to deal expertly with military matters. On the other hand the serving officers, mainly without previous intelligence experience, can handle the military technicalities but have steep learning curves in the business of foreign assessment.

In many ways these characteristics are sources of strength. The Cabinet Office system has the priceless assets of small size and informality. Because it is not self-sufficient it has every incentive to maintain its network of informal contacts in the Whitehall village. The wide area of consultation sucks in knowledge and builds up consensus, as does the engagement of operating departments such as the FCO and the Treasury. The preponderance of 'secondees' over professional intelligence officers also helps to minimise barriers between the intelligence and policy worlds. Moreover it enables the system to tap some of the best available talent, and injects plenty of first-hand knowledge of overseas affairs. These features enable it to hold its own with more sophisticated assessment systems like the American one.

On the other hand these strengths are also risks. The machinery is worked hard in keeping up with world events, and there is little double-banking to cover for weaknesses or learning periods. Since so much depends on those at the centre, variations in individuals' backgrounds and aptitudes have disproportionate effects. Its small size focusses the system on 'useful' and 'relevant' output, rather than more speculative items or apparently academic contingencies. The interdepartmental committee system is Rolls-Royce in its heavy claims on departments' time, and so also tends to be reserved for what seems significant; the Current Intelligence Groups would soon become discredited if they were used for speculative *tours d'horizon.* Few people stay in the central staff long enough to build up first-hand warning experience, and there is little institutional memory of past failures or knowledge of the academic literature about them. The balance of talent within the system also gives it something of a bias towards a 'reasonable' FCO view of the world, in which one lives with risk in the knowledge that the worst does not often happen. Since this is right for most of the time, the odds are somewhat stacked against a more neurotic world-view.

To trace the detailed effects on the handling of the Falklands question would need a separate study: all that is suggested here is that such details of organisation and staffing are significant in warning performance. Any organisation for national intelligence assessment involves balances between the conflicting *desiderata* outlined earlier, and the solutions adopted have their own characteristic effects and risks. Especially in dealing with small targets, quite small points of organisation and staffing can have considerable effects on outcomes.

This brings us back to the question posed earlier about the inevitability of warning failure. The academic writing on warning emphasises the general problems of cognition and the general pathologies of large-scale organisation, rather than more detailed organisational causes and effects of the sort just described. Of course the general factors apply, but it may be wondered whether they quite bring out the element of *contest* between intelligence and protective security. Surprise presupposes that the surpriser wins the contest between the defensive security screen covering his preparations and the 'offensive' intelligence collection of his potential victim. Misperception and organisational weaknesses exist on both sides. The surpriser can fail to perceive his security weaknesses, for the same reasons that lead the victim to mis-read the evidence of the surpriser's intentions. The biggest cognitive failure on the German side in World War II was, after all, in security rather than intelligence – in their dogged refusal to believe that the Enigma cipher machine could be vulnerable to cryptanalytic attack.[17]

There are of course characteristic outcomes of these intelligence-security contests. To give a military analogy, the particular terrain or technology may favour the defence against the offence: in our case, the surpriser against his victim's warning system.

For the last fifty years secretive regimes with effective security practices have undoubtedly tended to succeed with surprise attacks. But more open societies, the increased power of the world's media, and changes in intelligence and security technology may change the pattern. All is contingent and not predetermined, and 'micro' organisational factors such as those discussed here are among the contingencies on which outcomes depend. In commercial competition, small differences in the efficiency and emphasis of competing organisations lead to progressively increasing differences in performance. The same applies in the intelligence-security contest.

This partially answers the question about remediability. Intelligence is fallible, but so is security. As in war or commercial competition, success goes to the side making the fewer mistakes, or exploiting the other's mistakes better. There is no ideal warning system, just as there is no single 'best' form of organisational design. Any form of warning organisation has its characteristic strengths and weaknesses, but this does not mean that (over time) they cannot be changed.

Whether the lessons of 1982 have since been correctly applied to the British system would also be a matter for separate discussion. The only conclusion to be offered here is that the normal operation of the peacetime intelligence machinery does not keep it fully run-in for warning purposes. Surprise attacks are rare; their lessons are easily forgotten; most people get by without one happening on their watch. As long as such attacks are a possibility, someone needs to look periodically with a critical eye at the current intelligence machinery, searching for weak spots and means of improvement.

Some years ago, when 'warning' in NATO still seemed an important ingredient of the cold war, I wrote that

> because the intelligence machinery is permanently in operation, it does not lend itself as easily as other parts of the Alliance's military apparatus to training, exercises, and formal evaluation. Those in positions of responsibility tend to be more occupied with the substance of intelligence assessment than with increasing the effectiveness of the process. After major disasters, ritual organisational changes are made, but old habits tend to resurface with a fresh cast. Yet it is perhaps in pointing to the managerial factors that practical experience

> can make a distinctive contribution. . . . The practitioner will remember the details that academic and official history rarely record: the faulty distribution of a particular line of papers; the inexperience of a particular desk officer who had been left to sink or swim; the effects down the line of a particular chemistry of senior prejudices; and so on. The detailed working of the warning machinery is surprisingly important, and getting it right puts a particular onus on local management. An effective warning system needs some people in its supervisory chain with the qualities of good Regimental officers, ever-alert to dirty rifle-barrels or to slackness in the recesses of the cookhouse.[18]

No one suggests that any individual slackness was a factor in British intelligence performance up to April 1982. But the conclusion that the warning system should not be left to look after itself might be drawn from that experience, and could still be valid today.

NOTES

The author acknowledges support from the Leverhulme Trust.

1. F.H. Hinsley, 'British Intelligence in the Second World War', in C. Andrew and J. Noakes (eds) *Intelligence and International Relations 1900–45* (Exeter: University of Exeter, 1987) p. 218.
2. For surprise (and warning failure in general) see inter alia R. Betts, *Surprise Attack: Lessons for Defense Planning* (Washington: The Brookings Institution, 1982); M. Handel, 'Intelligence and the Problem of Strategic Surprise', in his *War, Strategy and Intelligence* (London: Cass, 1989); E. Kam, *Surprise Attack: The Victim's Perspective* (Cambridge, Mass.: Harvard University Press, 1988); and A. Levite, *Intelligence and Strategic Surprises* (New York: Columbia University Press, 1987). For a debate on the inevitability of surprise see Betts, 'Surprise, Scholasticism, and Strategy' (a critique of Levite's book), with a response by Levite, 'Intelligence and Strategic Surprises Revisited', in *International Studies Quarterly* 33 (1989) pp. 329–49.
3. James Rusbridger, *The Intelligence Game* (London: The Bodley Head, 1989), p. 2.
4. A.C. Maurer, M.D. Tunstall, J.M. Keagle (eds), 'General Introduction' to *Intelligence: Policy and Process* (Boulder: Westview, 1985) p. 6. This enthusiastic view of the role of intelligence is derived from Sherman

Kent's *Strategic Intelligence for American World Policy*, first published in 1949 (1965 edition, Hamden, Conn.: Archon Books, 1965).

5. Handel, 'Intelligence and the Problem of Strategic Surprise', p. 229.
6. For an introduction to deception see B. Whaley, *Codeword Barbarossa* (Cambridge, Mass.: MIT Press, 1973); and D.C. Daniel and K.L. Herbig (eds), *Strategic Military Deception* (New York: Pergamon, 1982). The statistics are quoted by R. Heuer, 'Cognitive Factors in Deception and Counter-Deception' in Daniel and Herbig, p. 60, and are taken from Whaley's *Strategic Deception and Surprise in War* (Cambridge, Mass.: MIT,1969), p. 164.
7. R. Jervis, *Perception and Misperception in International Politics* (Princeton, NY: Princeton University Press, 1976), p. 180.
8. Betts, *Surprise Attack*, p. 92.
9. T.L. Hughes, *The Fate of Facts in a World of Men* (New York: Foreign Policy Association, 1976), p. 24.
10. S. Gazit, 'Intelligence Estimates and the Policy-Maker', in M. Handel (ed.), *Leaders and Intelligence* (London: Cass, 1989), pp. 282–7. For discussion of the problem of intelligence's credibility with policy-makers and access to them, see also Handel's Introduction.
11. Conference paper by R. Jervis, *Strategic Intelligence and Effective Policy*, prepared for a conference on 'Security and Intelligence Needs for the 1990s: A Canadian Perspective', sponsored by the Canadian Association for Security and Intelligence Studies, Ottawa, September 1989.
12. R. Jervis, 'Intelligence and Foreign Policy'; *International Security 11* (1986–7) p. 161.
13. Command 8787, *Falkland Islands Review* (London: HMSO, 1983). The picture of the British organisation in 1982 is taken from the Committee's account, and from Sir Reginald Hibbert, 'Intelligence and Policy' in *Intelligence and National Security*, 5 (1990), pp. 110–27. See also M.E. Herman, 'Intelligence and Policy: A Comment', ibid.
14. For example Hibbert, 'Intelligence and Policy', p. 124.
15. Herman, 'Intelligence and Policy: A Comment'.
16. For the role of a central staff in an interdepartmental system, see E. Thomas 'The Evolution of the JIC System up to and during World War II' in C. Andrew and J. Noakes (eds), *Intelligence and International Relations* (Exeter: Exeter University, 1987), pp. 219–34; and D. McLachlan, *Room 39* (London: Weidenfeld and Nicolson, 1968), pp. 240–64.
17. For the German navy's examination of the evidence of cipher insecurity see T. Mulligan, 'The German Navy Evaluates Its Cryptographic Security', in *Military Affairs* 49 (1985) pp. 75–9. For details see F.H. Hinsley and others, *British Intelligence in the Second World War* Vol. II (London: HMSO, 1981), Appendices 1 (part ii) and 19; Vol. III part 2 (1984) pp. 51–2 and 230–1.
18. M.E. Herman, 'Warning: Some Practical Perspectives', in J.N. Merritt, R. Read, and R. Weissinger-Baylon (eds), *Crisis Decision-Making in the Atlantic Alliance: Perspectives on Deterrence* (Menlo Park, Ca.: Strategic Decisions Press, 1988), p. 7–3.

8 The Falklands War and the Concept of Escalation

Lawrence Freedman

In *Signals of War*, which I co-authored with Virginia Gamba-Stonehouse, we indicated that we had decided against drawing out the more general lessons that might be taken from our study.[1] My aim in this chapter is to consider the relevance of this experience to theories of limited war and crisis management. In particular it will focus on a general set of concerns normally expressed through the concept of escalation. I also intend to touch upon a couple of aspects related to nuclear weapons in order to warn that even in the most self-evidently non-nuclear crisis they can never be wholly excluded.

Many of the more obvious 'lessons' of the conflict (in the sense of stimulants to thought rather than rules to be applied) concern the special requirements of limited conflicts in a highly attentive political environment, both domestically and internationally. In conflicts of this type there are not many battles or even skirmishes. For long periods nothing happens, and then this is followed by bursts of intensive activity. Even the most limited encounter between the opposing forces can have important political consequences. A good example of this was the operation to re-take South Georgia. The operation was almost a disaster. An advance party was helicoptered onto a glacier on which it got stuck. Two helicopters crashed trying to rescue it but a third succeeded. These mishaps were kept secret, partly so that Argentina was unaware of the loss of those assets. On 25 April an Argentine submarine was observed reinforcing the garrison. The operation was immediately brought forward. The submarine, the *Santa Fe*, was severely damaged by missiles and depth charges from helicopters and was forced to ditch on land. Marines landed and surprised the garrison, which surrendered without much resistance, providing the desired impression of effortless British victory. If this first operation had ended as a fiasco then it could have finished the whole campaign, leaving a completely different impression to the one eventually left. Failure might weaken the bargaining position of the government *vis-à-vis* the opponent, but apparently provocative action can lead to diplomatic isolation.

The potential political significance of individual encounters creates particular problems for command and control. Arrangements for small-scale engagements might well be delegated to field commanders following standard operating procedures. Yet because these limited engagements can have major strategic implications, there is a case for a high-level political input. That, in turn, can lead to accusations of micro-management. Decisions by either local commanders or senior politicians will inevitably have to be taken on the basis of imperfect information, a problem that the local commanders may appreciate more than the politicians, especially when the intelligence involved is largely of a tactical nature.

However, not only do military commanders have to adapt but so do the decision-making processes back home. While there was something less than the future of civilization as we know it at stake in the South Atlantic, it was clear to both countries that the crisis was a 'threshold event' and that the consequences of either success or failure (but especially the latter) could be enormous. The stakes were sufficiently high to dominate government business for almost three months. The conflict therefore provides some insight into crisis decision-making in contemporary states. Inevitably there are profound variations in political institutions and culture which limit the conclusions that might be drawn for other countries.

For example, Britain's combination of highly centralised decision-making, backed by a professional civil service,contrasts with the more open and decentralised American system. Those responsible knew each other well, worked physically close together (the relevant ministries are within walking distance) and on the civilian side were professional civil servants with long experience rather than political appointees. Moreover no attempt was made to ensure that every service branch or agency of government had a fair share of the action according to some notional bureaucratic formula. One service – the Royal Navy – was given the operational lead and this simplified command and control. Because it never appeared as if the government was completely losing the military and diplomatic initiative, the political system was never put to a severe test. The government was backed in the essentials of its conduct of the war by the main opposition parties and by public opinion. The conduct of the war triggered neither an economic nor a diplomatic crisis. Allies stayed more or less loyal.

Another characteristic of British policy-making – which would be shared by other major powers – is that the political leaders involved now had little relevant experience of war. Of the War Cabinet, the Prime Minister and the Chairman of the Conservative Party (Cecil Parkinson) lacked *any* military experience. The Defence Secretary (John Nott) had been a Gurkha officer but had not been involved in a war. The Deputy Prime Minister (William Whitelaw) and the Foreign Secretary (Francis Pym) had fought in the Second World War. The experience of the Attorney General (Sir Michael Havers) as a war-time Naval Reserve officer was the most relevant. This meant that they had some understanding of what might be happening on the ground but no feel for military command, and so all lacked a confident base from which to assess and challenge military advice. In a way this was as true of the military junta in Argentina as it was of the British government, in that the major military engagements for the Argentine forces had been with local guerrillas, while the close military involvement in domestic politics left little time for the refinement of the traditional arts of strategy and command.

In the absence of relevant experience both sides were influenced – perhaps more than they were aware – by some of the ideas of contemporary strategic studies. In Argentina there had been considerable interest in the role of military action in getting issues high on to the political agenda (as the Egyptian President Anwar Sadat was believed to have done during the October 1973 war); in Britain there is evidence of the influence of the concept of escalation in both the deliberations of policy-makers and in public debate. The modest value of this concept as a tool for policy-makers was soon demonstrated in practice, which brings us to another possible 'lesson'. It is striking that even in the most apparently limited and contained of crises, matters soon become very complicated. The most simple of military operations against a well-armed opponent some distance from home becomes extremely complex. Unavoidable problems of imperfect intelligence and imperfect communications, plus the persistent failure of adversaries to do quite the things expected of them, mean that the worst of decisions can be taken for the best of reasons, or at least that the notorious 'fog of war' is influential throughout. The degree to which this is so in a comparatively limited encounter serves as a warning of the potential impact in a major confrontation where many separate encounters will be happening simultaneously.

It is for this reason that the concept of escalation deserves close examination. In the study of crisis management it is very much to the fore, often by way of representing the consequences of a failure of management, but also as the process to be exploited in securing national interests during a crisis while preventing matters from getting out of hand. Prescriptions for crisis management often assume the possibility of maintaining firm political control, even once military actions have begun. Crisis managers depend on the possibility of gearing military operations to the demands of diplomacy: a study of the actual processes of crisis management can help warn of the difficulties of achieving that end.

The concept of escalation was introduced in the late 1950s to describe the tragic process by which a limited war might turn into a total war.[2] It was soon appropriated by those who believed that the process by which a conflict grew in scope and intensity could be controlled and that success in manipulating this process could ensure a favourable outcome. This approach was closely linked to the search for the means of conducting a crisis and even a limited war with the Soviet Union without bringing about a nuclear holocaust. As such it was encouraged by the Cuban missile crisis of 1962. It also influenced US practice during the Vietnam War, especially as the Johnson administration sought to assess the alternative forms of pressure that might be brought to bear against Hanoi during 1964–5. 'Escalation' became a familiar term in explanations of what went wrong in US policies, and this meant that it began to regain its negative connotations. However the legacy of this period remains the idea of a process that can either be deliberate and controlled or else inadvertent and uncontrolled.

The key assumption behind crisis management is that if the crisis cannot be managed then the result will be war, or if fighting has already started then it will get much worse. The essential distinction between a managed and an unmanaged crisis is between controlled and uncontrolled escalation. There is a presumption in much of the literature on crisis management that crises are a recognised feature of international affairs, which require the adoption of a series of special procedures developed over time and refined with practice. Yet when one talks of standard crises they cease to be crises at all. Life has become a bit busier but there is no sense of impending disaster. A real crisis is when nobody can be sure that it is going to be managed. This is precisely because the problem has not appeared in an expected form and according to a recognised type and so the response has to be improvised. So it is important not to overstate the significance of standard procedures to manage crises; more important is a sensitivity to the variables.

Notions of escalation and graduated response create an expectation of conflict developing in discrete stages along a linear path. At least in the early stages it is assumed to be a negotiating process. This view understates the impact of operational factors on attempts to gear military action solely to the progress of negotiations. It also means that those entering into a conflict may have wholly erroneous expectations as to its future development. The dynamics of crisis and war ensure that the expectations of gradualism are rarely fulfilled. This can lead to mistakes, in terms of governments taking inappropriate measures of crisis management, and to accusations of bad faith in attempts to reach a settlement. Because it is now so often assumed that military moves in the early stages of a conflict are intended to carry a political message rather than simply reflect a prudent measure of preparedness, more can be read into these moves than is warranted. The more a conflict develops, the greater the danger of over-interpreting military moves as a series of diplomatic signals, for they are increasingly likely to be geared to operational imperatives. The concept of escalation has therefore become more influential in shaping attitudes towards crisis and war than its explanatory and predictive power warrants.

Although the concept of escalation was developed with nuclear crises very much in mind, it has been applied most often in non-nuclear crises. However, whenever a nuclear power is involved in a crisis there is inevitably some, albeit often quite marginal, nuclear dimension. Thus in the Falklands, while it was truly inconceivable that Britain's nuclear capability would be brought into play, its existence cast a small shadow over the proceedings. A nuclear power can never quite relieve itself of that status even in circumstances when it is in everyone's interests that it should do so.

The non-use of nuclear weapons – what Raymond Aron once described as the 'nuclear taboo' – has become part of the normative framework of international relations. This norm was given some legal standing in the form of Negative Security Assurances.[3] Even considering nuclear use would have seemed excessive and even preposterous in the circumstances of the South Atlantic conflict. Britain's nuclear capability was irrelevant to the outcome of the Falklands War. Even if Argentina had been a nuclear power, as it might be one day, it is by no means clear that nuclear capabilities would have played a significant role. At most the British nuclear capability would have provided some reassurance that nuclear weapons would not have been used against British forces. However Argentina would have had more to lose from nuclear engagements so close to its shores: it could not have been able to retaliate against Britain itself – unless it also developed a submarine-launched missile system – while it would not wish the islands at the centre of the dispute to become irradiated as a result of a nuclear detonation.

Nevertheless, it has been reported that the nuclear option was raised in internal studies, only then to be dismissed, as the British government began to organise its response to the Argentine invasion of the islands on 2 April 1982. A paper was prepared for ministers setting out the military options available to the government. The question of how far Britain should be prepared to go in a confrontation with Argentina was raised. It was suggested that the nuclear option should at least be addressed. It was concluded that *Polaris* submarine-launched missiles provided the only reliable nuclear option,[4] that this would require removal of a nuclear submarine (a SSBN) from NATO and that there was little chance of Britain exercising such an option. One source is quoted as saying that:

> Certainly the nuclear option was one of the options studied on 2 April . . . part of the work done that day involved examining the possibility of retargeting Polaris against Argentina.[5]

This was never taken seriously as a realistic possibility and thereafter no staff work was devoted to the matter. Nonetheless allegations were still made after the war – largely in the *New*

Statesman (the repository of many such allegations) – that a SSBN went as far south as Ascension Island in order to threaten or execute a demonstrative nuclear attack against Cordoba in northern Argentina. This was to be a response to the sinking of one of the capital ships – a carrier or a troopship such as *Canberra*. The decision to send the submarine was said to have been taken in the aftermath of the sinking of *HMS Sheffield* as ministers confronted the possibility that some more well-placed Exocets could spell defeat for the task force.

Paul Rogers has claimed that the submarine was located 'some hundreds of miles south-west of Ascension Island' and that there were insufficient hunter-killer submarines to meet the task force's requirements in the South Atlantic because two were held back to protect a *Polaris* SSBN.[6] This all seems unlikely. It is not normal practice for British *Polaris* submarines to have any escorts; it would have taken many days for a SSBN to reach the Ascension Island area from its normal patrol areas; it would not have been able to stay long before it would have been obliged to return to the UK in order not to exceed its normal patrol time; the Soviet Union would have been well out of range so there would have been a direct contradiction of official policy on the deterrent; it would also have contradicted those statements made by Admiral Sir Henry Leach and others at the time of the specific allegations, to the effect that there were no alterations to the standard patrol pattern. Furthermore, it remains unclear what possible strategic advantage the United Kingdom could have gained at any point in the conflict from the use of nuclear weapons.

There is one footnote to this issue. During the conflict public opinion was polled regularly. In terms of the prevailing views towards the conduct of the crisis it is worth noting as a general point that a substantial portion of opinion tended towards a more hawkish stance than that adopted by the government. This came out most clearly during April when a variety of options were still being canvassed before the implications of pursuing them had been fully aired. A poll undertaken by Market and Opinion Research International (MORI) on 14 April discovered some 28 per cent were prepared to bomb Argentine air and naval bases and 21 per cent willing to contemplate invading the Argentine mainland. Astonishingly, five per cent were ready to use nuclear weapons against Argentina.

Opposed to such a step were 93 per cent, with two per cent finding it difficult to make up their minds.[7]

Not long after the conflict was over there were a number of reports that the task force had carried nuclear weapons into the South Atlantic, not with the intention of using them but because they were part of the normal complement and had not been removed. The allegations were based on reports that a deep-diving vessel had been sent to recover equipment from the sunken wreck of *HMS Coventry*, a Type-42 destroyer. Earlier reports had spoken of attempts to find nuclear weapons among the wreck of *HMS Sheffield*, another Type-42 destroyer.[8] However, it is now believed that the deep-diving vessel was attempting to recover top-secret cryptographic equipment and codebooks. In any case, it is not normal for Type-42 destroyers to carry nuclear weapons.[9]

George Quester has claimed that British conduct during the war created 'dangerous nuclear precedents', but this was largely on the basis of accepting the rumours that British ships in the combat zone had nuclear weapons on board as well as the use of nuclear-powered submarines. Given the anxiety over the anti-submarine problem faced by Britain, Quester asks whether there would have been a temptation to have available to the fleet the full range of anti-submarine capabilities and even to use nuclear depth charges in the event that the Argentine submarines performed better than turned out to be the case. He notes that this might have been rationalised on the basis that there would have been no collateral damage to civilians and no radioactive fallout.[10] However, all this is wholly speculative as the British government showed no inclination to succumb to such temptations.

Normally British carriers (in this case *HMS Invincible* and *HMS Hermes*) and some frigates carry nuclear depth charges for anti-submarine warfare purposes. There seems to be little doubt that some ships, especially those that came straight from exercising off Gibraltar, were carrying nuclear weapons. It has been reported that three-quarters of the total naval stockpile set off towards the South Atlantic – but also that ministers were horrified when this was discovered and ordered a Royal Fleet Auxiliary to collect them at Ascension Island. It has also been suggested that after this discovery there was great reluctance to let nuclear weapons go to sea again for some time.[11] Despite the fact that

Britain *did not* take nuclear weapons into the South Atlantic the allegation that it did, and that some nuclear weapons were lost when ships were sunk (and are thus now possibly polluting the South Atlantic) still exercises a powerful hold on some imaginations. Officials in the Soviet embassy in Buenos Aires were reported as spreading the charge of nuclear contamination, citing as evidence the appearance of a number of dead penguins, and warning that 'the Argentines have their own Chernobyl in the Atlantic'.[12]

The final link between British nuclear capability and the South Atlantic is the less deniable presence of nuclear-powered submarines during the conflict and since. Britain's nuclear-powered submarines were seen to exercise a critical influence on the course of the conflict, especially but not solely by the sinking of the Argentine cruiser, the *General Belgrano*, by *HMS Conqueror*. This was not in itself a violation of the Treaty of Tlateloco,[13] a point acknowledged by Argentina.

Having reviewed the various claims with regard to British nuclear conduct during the Falklands War and the likely Argentine response, the general conclusion must be that the nuclear dimension was minimal. However, what is perhaps significant is not that these claims lacked substance but that they were made at all, and have continued to surface despite a lack of evidence to sustain them. This indicates that nuclear powers cannot escape their status even when it has slight relevance to a particular conflict. If a nuclear power is involved in a military confrontation then a nuclear dimension is acquired immediately. For navies the risks connected with the carriage of nuclear weapons into whatever combat zone they happen to be entering even when there is no expectation of a clash with another nuclear power are important to keep in mind. Significant damage to a ship carrying nuclear weapons could turn a serious incident into a major crisis with considerable political repercussions. It might be argued that the United States has been involved in a number of crises without the same attention being given to the fact that it is a nuclear power. One reason for this may be that the US has much more conventional capability at its disposal than Britain. The anxiety that gave the various concerns about British nuclear capability any credibility at all was that the war was going to be a very close call and that Britain might find itself in an embarrassing and fraught situation, perhaps as the result of the loss of a carrier or troopship. In these circumstances there could be an attempt to escalate in order to regain the strategic initiative and the obvious instrument of escalation would be nuclear weapons.

My own view is that the 'nuclear taboo' is sufficiently strong to cope with such pressures and that there was never any risk that Britain would move in that direction. However, the sensitivity of nuclear issues is such that they will be raised, both domestically and externally, even when in practical terms there seems little reason to do so. What is believed to be the case with regard to nuclear risks is more likely to be important than what is the case. Any risk of nuclear escalation is considered unacceptable and a nuclear power can expect to be charged with running such a risk – almost by virtue of being a nuclear power.

The importance of popular and even elite attitudes when it comes to shaping crisis behaviour is also apparent when considering the approach to escalation in general. In low-intensity conflicts it is assumed that any escalation will be deliberate and controlled. This creates certain expectations. The first is that conflicts will develop in a gradual and incremental fashion, and the second is that each step change should reflect a considered political judgement geared to clear political objectives. The influence of these ideas can be illustrated both in the crisis which led to the war itself and in the later management of the war. We will examine them by looking at two of the critical events of 1982 – the March crisis which concluded with the Argentine decision to invade and the sinking of the *Belgrano* in early May.

Before doing so it is important to make two points with regard to the policy-making process and escalation. It is often stated that one way of keeping limited conflicts limited is to set limited objectives. This reflects the view that there is naturally some proportionality between ends and means. Unfortunately, as always, the matter is not so simple. Certainly in the early stages of a conflict governments will be aware that some objectives cannot be met with the available means or that, if they could be, the relevant means would be inappropriate and their use would create more problems than they would solve. However, once a government has committed itself to certain objectives, even if quite limited, then its stake in achieving these objectives goes beyond their intrinsic worth. In part this is a matter of international prestige and reputation. There are obvious dangers in allowing the demonstration of 'resolve' to become an overriding objective, but most governments are keenly aware that their international image depends on their ability to handle major crises and that visible scaling down – or dramatically failing to achieve – the objectives can damage this image, perhaps more than it is bolstered by success. Even more seriously their position at home will be jeopardised by failure. The Conservative government was not popular at home in early 1982 and there was a widespread belief that the Prime Minister at least could be forced to resign if the task force failed. The successful prosecution of the war saw a major surge in the popularity of the government.[14] In Argentina the junta was obliged to leave office following its defeat, joining a long line of dictatorships (Portugal, Greece) to have fallen in the aftermath of military failure.

Although it is often assumed that in modern welfare states there is no appetite for the defence of interests through the application of armed force, the evidence from this conflict is that public opinion can be highly supportive. Thus while it was the case that up to a quarter of the British electorate were anxious with regard to the human and material costs of the conflict and opposed provocation, an equal number wanted the government to take tougher action. Furthermore, as the conflict developed there was a growing readiness to accept loss of life to achieve the removal of Argentina from the islands. When British sailors and soldiers died this was attributed by the majority of the public to Argentine intransigence which, along with the original aggression, ought not to be rewarded.[15] A conflict can create its own stake in the outcome, over and above the original interest. In the case of the Falklands, because there was no influential body of opinion challenging the government's stance, and no major disaster in the conduct of the campaign prompted second thoughts, one has to be careful in making too much of this. Nonetheless both governments were very aware of public opinion as a factor driving them on as much as holding them back. Both stressed the strength of popular feeling during the early stages of the meditation effort of US Secretary of State Alexander Haig so that he could realise the limits to their room for manoeuvre.[16] When it came to constructing possible compromises it was difficult to prepare public opinion for any climb down. Insisting to a domestic audience that no serious concessions were being made helped increase the suspicions on the other side that this was indeed the case.

British policy towards the Falkland Islands up to March 1982 was shaped by tension between the logic of the geo-political situation which favoured Argentina, and a series of past commitments with regard to self-determination, backed by an effective Falkland Islands Lobby and supported by Parliament, which favoured the islanders. As the two positions became polarised, the government could only hope that in the long-term common sense would prevail and the two sides could be reconciled. But reconciliation was only likely if Britain put pressure on the islanders to negotiate. The moment never seemed appropriate for such pressure. However, although there was a lack of political will in London to solve the dispute once and for all by some deal with Buenos Aires, there was no inclination to accept full responsibility for the long-term security and prosperity of the islands.

In June 1981 in the course of a Defence Review it was decided to scrap the ice-patrol ship *HMS Endurance* – the sole regular British naval presence in the South Atlantic which had taken on a symbolic importance far beyond its military capabilities. The Foreign Office warned – correctly – that this could well be misread in Buenos Aires. This would leave a small garrison of Royal Marines to deter Argentina from attempting to retake the Falkland Islands by force. The British government was weakening its position at a time when the islanders and the Argentines were hardening theirs.

In Buenos Aires, patience had run out with Britain. The 150th anniversary of the British seizure of the islands in January 1983 appeared as a sort of deadline. The government of General Leopoldo Galtieri, which had come to power in December 1981, had the issue high on its agenda. Planning for a possible invasion had begun in January 1982. The Argentine government noted precedents (such as the Indian takeover of the Portuguese colony of Goa in 1961) where after initial condemnation the international community had learned to accept the results of decisive military action. In Washington, the Galtieri regime was judged to represent the acceptable face of military dictatorship. Cooperation was developing on the support of other right-wing regimes in Central America. The hope was that Washington would not be too cross if Las Malvinas were retrieved. The timing of the Argentine invasion indicates a lack of concern for minimising Britain's ability to respond and indeed Argentine leaders have since confirmed that they assumed Britain would not respond. An invasion was not scheduled for the start of April. The fact that it took place then was the product of the events of March 1982.

The trigger for the crisis came when Argentine scrap metal merchants led by Snr. Davidoff landed on South Georgia on 19 March. Davidoff had a contract to take materials from Leith but no means of doing so. The Argentine navy was happy to help him do this so as to be able to establish a presence which might continue should the British Antarctic Survey leave in October 1982.[17] There was no need for this presence to be obtained surreptitiously. The arrangement had been agreed with the British Embassy in Buenos Aires. However, the naval ship taking the scrap metal men to the islands failed to follow the normal formalities, in particular by not paying a courtesy call on the British Antarctic Survey base at Grytviken. If there had not been a developing tension in Anglo-Argentine relations at this point, this incident might have passed off with a mild official protest by Britain. Such was the instinct of the British Embassy in Buenos Aires. However, others were more suspicious. In 1977, Argentina had established a presence in an even remoter dependency – South Thule – and there was concern that the plan was now to do likewise on South Georgia.[18] There had been a similar incident the previous December and events now seemed to be following a pattern. The government's critical decision was to demand, not just that the Argentines abide by the rules, but that they should leave altogether.

The instinctive desire of the Foreign Office to sort the matter out through private diplomacy was undermined by the news of the 'landing', and the accompanying raising of the Argentine flag, leaking in the British press. This led to demands in Parliament for a reassuring statement. The British also felt themselves misled, believing they had been told that the scrap metal merchants would leave with the ship that brought them. It was then discovered that they had stayed behind, although the record is unclear on this point. As they could no longer leave under their own steam, they would have to be removed by *HMS Endurance.*

The British government refused to acquiesce in any 'infringement of British sovereignty' but there was little it could do to back its position by force, except dispatch *HMS Endurance* to South Georgia. The Prime Minister observed to the Commons on 3 April,

> had I come to the House at that time and said that we had a problem on South Georgia with 10 people who had landed with a contract to remove a whaling station, and I had gone on to say that we should send HMS "Invincible", I should have been accused of war mongering and sabre rattling.[19]

The government became convinced that sovereign British territory was about to be seized by a foreign power just two days before it happened, leaving time only for frantic but futile diplomatic activity. Up to this point there had been some Cassandras in the intelligence community but the general view was disbelief that the Argentine would take such a drastic step. There was a greater readiness to believe in a campaign of graduated pressure. The critique made by the Franks Committee, set up to examine why Britain had been surprised by Argentina, is of interest here. The Foreign Office, it suggested, had underestimated the importance attached by Argentina to solving the dispute in 1982, had been unduly influenced by a previous history of bellicose noises from Argentina not being followed up by serious action, and thirdly, 'in believing on the basis of evidence, that Argentina would follow an orderly progression in escalating the dispute, starting with economic and diplomatic measures'.[20] Had Argentina adopted such an approach it might actually have been in a better position: Britain would have been put on the spot without being provoked into a major military response. (It did not because it feared that this would encourage British cooperation with Argentine's main local rival, Chile.) The Franks Report suggests that the presumption of gradualism did not make 'sufficient allowance for the possibility of Argentina's military government, subject to internal political and economic pressures, acting unpredictably if at any time they became frustrated at the course of negotiations'. This suggests that the aggravating factor which turned Argentine frustration over British intransigence into such drastic action was the domestic state of the country. However, the aggravating factor was more in Britain's attempt to convey an impression of firmness that was not in practice backed up by any action.

In attempting on 23 March to reassure Parliament and the Falkland Islanders, the Foreign Office Minister of State Richard Luce alarmed the Argentines. The despatch of *HMS Endurance* to South Georgia, apparently to expel the scrap metal merchants, brought national honour in to play. Argentine warships were deployed, in order to warn that such an expulsion could escalate. Hints that a nuclear submarine and even frigates were being sent by Britain to the South Atlantic led to concerns that the particular incident on South Georgia was being used as a pretext by London to reinforce its military position in the area. If true this would complicate and perhaps rule out any future plans, which were still being drawn up, to take the islands by force. On 26 March the junta decided to strike while the opportunity was still available.[21]

In terms of strategic studies this is almost a text-book example of pre-emption. One of the many disturbing features of this decision was that the intelligence appreciation upon which it was based appears largely to have been derived from compilations of press comment from Britain, much of it highly speculative. As always the hardest thing to anticipate is the heroic misunderstanding by the adversary. At a time when Foreign Office ministers were deciding *not* to send some ships which had been exercising off Gibraltar to the South Atlantic because this would be too provocative, the Argentine Ministry of Foreign Affairs was convinced that these ships *must be* on their way because of press reports that they *could be* on their way. Statements by ministers to Parliament which were designed to sound tough without actually saying anything may have sounded more impressive in Buenos Aires than in London. The Argentine action thus represented a deliberate act of escalation, but one taken in the belief that the British government had already taken their own steps towards escalation. Models of deliberate escalation often assume that one side is continually taking the initiative – upping the ante in order to extract political concessions from the other side. The junta in Buenos Aires believed that if it did not act its bargaining position would worsen as a result of the reinforcement of the British presence in the South Atlantic. This case suggests that the dynamics of escalation are likely to be fuelled by essentially defensive rather than offensive instincts and that the hardest thing to appreciate during the course of a crisis is a comprehensive misunderstanding of one's motives and behaviour by the adversary.

In seeking to regain the Falklands, the major military problem was time. According to Lord Lewin, the then Chief of the Defence Staff, this was because of the sustainability of the task force and the onset of the Antarctic winter.

> The best estimate that could be given to Ministers was that the Task Force could sustain operations for a maximum of six months from the time of sailing, and for the last two or three months of that time it would be likely to become increasingly less effective.

The logistic problem was compounded by the meteorological one. In July, three months after sailing, it would be midwinter and land and sea operations would become increasingly hazardous. It would be difficult to keep troops at sea without them becoming demoralised and unfit.[22] The Task Force could not stay in the South Atlantic indefinitely.

The military instinct was therefore to concentrate from the start on preparing for a landing on the islands. The need for special forces to be inserted on the islands to build up a picture of the Argentine land forces meant that it was necessary to get close to the islands regularly. The main concern was that the Argentine air force and navy should be engaged prior to a landing so as to cause the maximum attrition and therefore reduce the opposition to the eventual landing.

The first stage of the military operation came on 1 May. With a considerable effort a *Vulcan* bomber attacked Stanley airport early in the morning. On the same day there were a number of Harrier raids and naval bombardments against Argentine positions on the islands. This was designed to give the impression of an attempted landing, so as to draw out the Argentine navy and air force.[23] In particular it was hoped to attack the Argentine carrier the *25 de Mayo*. The War Cabinet had given permission for a submarine to attack the carrier should it cross its patrol area as it got itself in position to launch air strikes against the task force. However, the submarine failed to find the carrier, while the carrier failed to launch its aircraft. At the end of the day the Argentine commanders concluded that no British landings were taking place and that prudence required a return to safer waters. The British commanders did not know this. They were worried by their failure to find the carrier which they assumed to be closing on them. The only major Argentine ship that they had found was the old cruiser the *General Belgrano*, escorted by two destroyers, and trailed by another submarine, *HMS Conqueror*. The task force commanders wanted to attack this target but had to wait for a political decision to permit such an attack. By the time the decision could be taken the next day, the cruiser had turned round and was heading away from the task force. On the assumption that it was still closing, ministers authorised an attack on the *Belgrano*. *HMS Conqueror* received the order authorising the attack as it was transmitting its report of the change of course. The cruiser was torpedoed with the loss of 360 men in the most costly single engagement of the war. All Argentina could do was to avenge the loss of the *General Belgrano*. The destroyer *HMS Sheffield* was surprised on 4 May by an air-launched Exocet missile. Twenty sailors were killed, and the ship abandoned to sink. The British acted in the belief that the *Belgrano* was part of a general Argentine offensive against the Task Force. Such a general attack had indeed been ordered but it had been called off. The intelligence appreciation upon which the government agreed to sink the boat was simply out of date.[24]

During that same day, 2 May, some progress had been made in Peruvian-Argentine discussions of a revised and simplified version of Haig's peace plan. President Belaunde of Peru assumed that Secretary of State Haig was doing the same with British Foreign Secretary Francis Pym who was in Washington at the time, and it was assumed in Peru (and so in Argentina) that when Haig spoke he was virtually speaking for Britain. In fact Haig made slight progress with Pym who was expecting the next diplomatic effort to be led by UN Secretary-General Perez de Cuellar. Haig confused both sides. In the evening of 2 May President Galtieri was to put the Peruvian plan, which had yet to be discussed seriously with the British, to the rest of the junta. News of the sinking of the cruiser arrived, leading to the rejection of the Peruvian proposals.

At the time, the main diplomatic cost was not seen in the loss of the Peruvian initiative – which was actually sustained until 7 May – but in the loss of international political support for

Britain's case. The victim was just outside the 200-mile exclusion zone. Although Britain had been very careful not to suggest that this was a combat-inclusive zone it had been widely understood as such. Such a dramatic transformation of the crisis led to accusations of unwarranted escalation. Haig made known to Britain his concern that action such as this was alienating Latin America and threatening the Western position on the continent.[25] Many in the international community became distinctly uneasy about continuing with unequivocal support for Britain and more determined than ever to negotiate a cease-fire. Britain maintained its position, largely because for much of the first half of May it did seem to be taking attempts to reach a negotiated solution seriously. After the shock of the *Belgrano* and the *Sheffield*, the government adopted a more conciliatory attitude than it had been prepared to adopt before or was prepared to adopt later. On 6 May it accepted a version of the Peruvian proposals which reaffirmed the importance of the wishes of the islanders but did accept an interim administration made up of a small group of countries excluding Britain to supervise the withdrawal of Argentine forces and be involved in negotiations for a 'definitive agreement on the status of the islands'. This was the most that the British government was ever prepared to offer, but by this time Argentina was no longer interested.

The sinking of the *Belgrano* became a *cause célèbre* in Britain after the war. It became so because Secretary of Defence John Nott's original report of the action suggested that the Argentine cruiser had been sunk because it had been closing in on the task force; it soon became clear that this was not the case and that at the time of the sinking it had been turning for home. The critics have put a lot of energy into demonstrating that the government *must* have known of a promising possibility for a peace settlement, just as they *must* have known that the *Belgrano* was not a threat because of the direction in which it was sailing at the time it was sunk.[26]

The importance of this post-war debate is that it throws into relief common assumptions which are often adopted by politicians concerning the control of escalation. A military logic is expected to be subservient to a political logic. This points to a graduated response with each military escalation only justified if political remedies continue to be frustrated. All military action – at least in the early stages of a conflict while there remains hope of a political settlement – is expected to support diplomatic purposes. At work is what might be described as a theory of *Just Escalation.* The concept of just escalation represents an attempt to prescribe military conduct when hostilities are underway and the original wrong has yet to be righted. It has a resonance in traditional concepts of just war, contemporary strategic theory and public debate, as exemplified in the concept of flexible response. The theory of just escalation insists that any step change in the degree of armed force employed must be geared to the developing political situation and no more than that necessary to induce that change in the opponent's stance necessary to make possible an honourable settlement. *Unjust Escalation* by contrast involves excessive force and tends to be counter-productive – assuming that this is another of those liberal theories in which the ethical and the prudential are happily inter-twined. Unjust escalation reduces the chance of settlement and leads to a vicious and costly fight to the finish. Both are theories of controlled escalation.

These ideas find an echo in contemporary strategic theory in which it is often suggested that the ideal is for all military operations to be under firm political control – for the only purposes for which they can be used legitimately are political. In particular, any move beyond the existing limits to a conflict – the process known as escalation – should be taken at the highest political level and only after the most careful consideration of the likely military and political consequences. More seriously, in limited wars it is assumed that control will be exercised so as to reinforce a negotiating position.

The sinking of the *Belgrano* has come to be presented as a classic case of unjust escalation. The most extreme set of allegations suggests that the government knowingly acted in such a way as to preclude a diplomatic solution in preference to a military solution. In the more modest version, the War Cabinet pursued a military option recklessly, thereby losing and even wrecking – perhaps inadvertently – a chance for peace. In both versions the moral is reinforced by the sinking of *HMS Sheffield* on 4 May, which is seen as a form of retribution (although in practice the Argentines had been trying to sink British ships before the loss of the *Belgrano*). Yet the complex interactions between the armed forces of the two sides, and the relationship of military to diplomatic activity, make apparent the difficulty of relating military force to diplomacy in a controlled manner. This relationship is far more complex even in a limited war than much strategic theory might lead us to suppose.

There was no formal state of war between Britain and Argentina. Such a state would have carried with it many awkward implications for Britain, Argentina and other important nations. For example, the United States would have been required to declare its neutrality. The lack of a declaration of war meant that Britain was obliged to justify Operation Corporate in terms of the 'inherent right of self-defence under Article 51 of the UN Charter'. This justification was reinforced by Security Council Resolution 502 of 3 April 1982 which required Argentina to remove its forces from the Falklands.[27] Does acting in self-defence impose any limitations on the sort of military action permitted? At what point might it exceed the requirements of self-defence? The difficulty is that the exercise of the inherent right of self-defence *after* the seizure of territory by another country is likely to require the use of sufficient force to eject the enemy from that territory. If one is forced to stay on the tactical defensive, in respect of some concept of proportionality, then all the advantages flow to the aggressor whose offensive action has been completed. Thus what is left uncertain by the idea of taking measures in self-defence is the circumstances in which it is permitted to go on to the offensive.

There is therefore no reason in principle why self-defence cannot involve going on to the offensive. The commander of the Task Force was charged with bringing about the withdrawal of the Argentine forces from the Falkland Islands and re-establishing British administration there with the minimum loss of life. The problem demonstrated by the *Belgrano* episode is that once military operations have begun in earnest the question of what constitutes minimum force becomes moot and the casualties likely to result from any given operation are almost impossible to calculate.

Political constraints were put on military operations. Direct attacks on the Argentine mainland were advocated but ruled out because of the possible impact on international opinion (and also because of doubts as to their effectiveness). There was a continuing restriction on the conduct of any operations against Argentine territory or within Argentine territorial waters. The military at different times argued that these restrictions did impose penalties – they gave warships safe haven and aircraft safe bases. Nevertheless, because the British did not make these restrictions explicit Argentine commanders could not rely upon them. This uncertainty complicated Argentine military planning. Equally, Argentina decided not to attempt to attack passenger liners serving as troop ships, but the thought that it might was a source of great anxiety to the British commanders. Thus while both sides may accept that certain categories of military action would incur severe political penalties, they can gain at least some of the military advantages by not making this explicit. This means that the actual limits on warfare are always to a degree ambiguous.

The situation was similar with regard to nuclear submarines, Britain's major military advantage, but one which was difficult to use against merchantmen without breaking agreed rules of naval engagement. But the position against warships was more ambiguous. The politicians were gradually persuaded to overcome their inhibitions against using these submarines – first within the Exclusion Zone, then against the carrier and eventually against any warship outside territorial waters. This persuasion became more effective as it was felt that the limitations on the conflict were generally becoming eroded. But this means that there was no clear sense of the 'rules of the game' and that steps which went beyond what might have been understood as a threshold, taken for whatever reason, raised the political temperature. Although Britain had made statements to the effect that the Exclusion Zone by no means precluded combat outside, it was not surprising that enemy ships presumed that they would be safer outside it than in.

Thus, while the military were not allowed simply to defeat the enemy armed forces through whatever appeared to be the most expeditious and effective means, this did not mean that the limits within which force was being applied were evident to the other side. In order to gain the operational advantages that come with keeping the adversary guessing, Britain's apparent rules of engagement as notified through the United Nations to Argentina were less severe than those actually in use. Moreover, even when it is intended to give the impression that only measured force is being applied, in practice it is not always so easy to control the use of armed force. This is especially so when facing a competent and well-equipped enemy against whom it might be too dangerous to take risks. While it might be advisable for the sake of public support at home and abroad to appear moderate, it is also necessary to convince the adversary of resolve. An effective display of force might extract sufficient concessions in negotiations to make further and more violent action unnecessary. Because of these considerations the military found themselves under pressure to be mindful of the immediate diplomatic impact of their operations as well as their ultimate objective of retaking the islands. This reflected the view that, because such wars are settled diplomatically, military action must be geared to achieving a favourable eventual settlement. Action of this sort may be described in terms of 'force being the only language they understand', the need to 'turn the screw, teach a lesson, concentrate their minds', and so on.

With each turn of the screw there is a need to inquire as to whether the victim is becoming more responsive before turning again. But the British position in the South Atlantic did not allow for the gradualism that this implies. Time was running out. The need to prepare for an amphibious landing imposed its own strict timetable. The fear was that diplomacy would be used to reinforce the Argentina military position by disrupting this schedule. Argentina for its part was trying to hold on to a position – and therefore had a much clearer interest in a ceasefire. In March, however, it was Argentina which had eschewed gradualism,on the grounds that if it waited too long the British would have reinforced their naval presence in the South Atlantic to preclude any military operations to take the islands in the future. To some extent gradualism is a luxury that only those confident of their superiority and able to be patient can afford. For those fearful of defeat all restrictions imposed for political reasons – whether they be the risk of offending international opinion or the need to allow diplomacy to take its course – impose operational penalties.

The unavoidable tension between the political and military logic in the conduct of any limited war warns against any attempt to relate diplomatic action too closely with military action, simply because of the speed with which operational circumstances can change. If tactical decisions become too infused with political meaning then there is a risk that the outcome of particular engagements, which can never be exactly foreseen, will be wrongly interpreted. Gradualism itself is unlikely to be sustained. Military campaigns rarely involve a buildup to a grand finale; the bloodiest and most difficult confrontations can be among the earliest followed by more sporadic actions while both sides regain their strength. The rhythm and tempo of conflict will be influenced by much more than the rhythm and tempo of diplomacy. In limited war individual engagements can take on a great significance, yet all that we know about warfare tells us the nature and outcome of these engagements will be influenced by chance and uncertainty as much as by careful staff work and skill in combat. It will be necessary to operate with imperfect intelligence and poor communications. Moreover when this familiar 'fog of war' descends over a battlefield, it can soon fog any associated diplomacy. In addition, as the 'chinese whispers' from Haig through Belaunde to Galtieri testify, diplomatic activity can itself mislead and confuse.

Controlling the transition from crisis to war depends on military action being adjusted throughout to the requirements of diplomacy. However, unless the practicalities of military power are understood then it will not prove to be an adequate tool of diplomacy but will create diplomatic imperatives of its own. The military logic will drive a political logic. Once invoked, military power transforms diplomacy as compulsion takes over from compromise. Previously acceptable solutions become an insult to the men that died.

There is, of course, nothing new in the identification of the 'fog of war' – or its half-brother the 'fog of diplomacy' – as familiar characters in international dramas. Yet the unpredictable and uncertain is often played down in peacetime planning, in which it is often supposed that a course of action can be laid down for every eventuality. Once a major crisis breaks then the fog soon descends and the old lessons are relearned, but the learning process can take time and awful things may happen during that time. With the limited experience of most political leaders in high-level crisis management the learning process may now take longer than in previous eras, when a steady amount of great-power conflict was the norm. It is at the very least important that any crisis simulations designed to familiarise policy-makers with the procedures, stresses and strains of this sort of activity include a good measure of the unexpected, the incompetent and the misunderstood and do not over-encourage the notion of an orderly and systematic process.

It is important to recognise that even smallscale military encounters can have decisive political consequences. One of the gaps in British command and control arrangements was a level of command between the tactical field commanders, with their responsibility for specific tasks, and the headquarters at home, which set the tasks according to available information and government decision. The lack of this level meant that there was often a lack of clear overview of the military situation from close at hand, and of a buffer between the field commanders and the political pressures emanating from London. It might be added in this regard that another of the educational efforts required is in the area of logistics, which is often imperfectly understood in military commands and when not pre-organised to allow military exercises to run smoothly unavoidably slows everything down – much to the exasperation of politicians anxious for results.

Restraint in the use of force is much more feasible for countries with significant local military superiority, or at least with substantial reserves and a range of options that can be sustained over time. Countries with limited military power lack the diplomatic flexibility of a superpower. Still, any country operating far from home might note that military options cannot be maintained indefinitely and some are highly perishable, and if they deteriorate so does a bargaining position. This again often depends on the quality of logistical arrangements.

What do we expect of political leaders and their military and diplomatic advisers in these circumstances? They must ensure that they are privy to the innermost thoughts of the enemy, and clairvoyant about decisions that it has yet to make; follow, decode and correctly interpret enemy military orders, however complex the sequences and cryptic the message; make sense of complex and fast-moving events, in a variety of capital cities as well as theatres of war; identify the full operational and diplomatic implications of these events; appraise the available choices and the possible consequences of each; take a firm decision and then communicate it swiftly and unambiguously to those responsible for its implementation – who will then carry out their orders in such a way as to ensure exactly the result intended. All this would require as much good luck as good management. Events generally move faster than the capacity of decision-makers to absorb them and assess their meaning. Information will be imperfect, incomplete and often contradictory. The assumptions with which decision-makers fill the gaps may be hopelessly wrong yet perfectly reasonable. Political control over military operations is necessary but extraordinarily difficult. Military force is not a light and sensitive instrument. It is heavy and unwieldy, and can produce unexpected results.

NOTES

1. Lawrence Freedman and Virginia Gamba-Stonehouse, *Signals of War: The Falklands Conflict of 1982* (London: Faber, 1990) p. xxxvi.
2. For a discussion of the concept of escalation see Lawrence Freedman, 'On the Tiger's back: The Development of the Concept of Escalation', in Roman Kolkowicz (ed.) *The Logic of Nuclear Terror* (London: Allen & Unwin, 1987).
3. Britain has committed itself *not* to attack non-nuclear states through such an Assurance made at the UN Special Session on Disarmament in 1978. There the Head of the UK Delegation made the following statement:

   > I accordingly give the following assurances on behalf of my Government to non-nuclear weapon States which are parties to the Non-Proliferation Treaty or other internationally binding commitments not to manufacture or acquire nuclear explosive devices: Britain undertakes not to use nuclear weapons against such states except in the case of an attack on the United Kingdom, its dependent territories, its armed forces or its allies by such a state in association or in alliance with a nuclear weapons State.

   The United Nations Special Session on Disarmament, Cmnd 7267 (London: HMSO, 1978) Annex Q.
4. In the event Britain's other main nuclear delivery vehicle at the time, the Vulcan bomber, was used in raids on Port Stanley, although carrying only conventional weapons.
5. *New Statesman*, 31 August 1984.
6. The Belgrano Action Group, *The Unnecessary War* (London: Spokesman Books, 1988).
7. One of those responsible for collecting and interpreting this information attended a dinner party and expressed his horror that one-in-twenty Britons could take such a position. He was fixed with a frosty glare from an upper-class lady who informed the assembled company that 'If we have to nuke Rio to regain the Falklands, we'll do it'. Whether or not the lady was aware that Rio was not in Argentina, her point still held in that she saw no obvious limits to British action. The

poll was conducted on 14 April 1982 and was based on 1018 adults aged 18+ interviewed across fifty-three constituency sampling points across Great Britain. The results were published in *The Economist*, 16 April 1988. In a book containing his posthumously published letters, David Tinker (on *HMS Glamorgan*) quotes a colleague in early April suggesting 'Drop a big white job (Polaris) on them'. Tinker adds: 'Thank goodness he's not in command'. David Tinker, *A Message from the Falklands* (London: Junction Books, 1982) p. 158.

8. *New York Times*, 1 July 1988. Some dummy nuclear weapons, which are used for training purposes, may have been seen and mistaken for the real thing. Tinker reports on 5 June 1982:

> One of our jobs out here is to transfer stores around and between ships and yesterday I walked into the hangar and found a nuclear bomb there. . . . Of course, it turned out to be a drill round, full of concrete, that *Fort Austin*, now eventually going home, was taking back to England.

   Tinker, p. 198.

9. Duncan Campbell, 'Too Few Bombs To Go Round', *New Statesman*, 29 November 1985.
10. George Quester, 'The Nuclear Implications of the South Atlantic War', in R.B. Byers (ed.), *The Denuclearization of the Oceans* (London: Croom Helm, 1986).
11. Duncan Campbell, op.cit.
12. *Arms Control Reporter*, 23 June 1986.
13. At the time of the conflict the British Ambassador to the United Nations, Sir Anthony Parsons, wrote to President of the General Assembly that the treaty excludes 'an instrument that may be used for the transport or propulsion of the device is not included in this definition if it is separable from the device and not an indivisible part thereof'. Letter to President of the General Assembly A/S-12/29, cited in *Arms Control Reporter*, 15 June 1982.
14. It remains a matter for debate as to whether the 'Falklands factor' was a major determinant of the sweeping Conservative victory in the General Election of May 1983. For a powerful argument to the effect that it was not see David Sanders *et al.*, 'Government popularity and the Falklands War: A reassessment', *British Journal of Political Science*, 17 (1987), pp. 281–313.
15. For a discussion of opinion poll evidence in Britain, see Lawrence Freedman, *Britain and the Falklands War* (Oxford: Blackwell, 1984) Ch. 8.
16. See Alexander Haig, *Caveat* (London: Weidenfeld and Nicolson, 1984). Mrs Thatcher's method was to point out pictures of Nelson and Wellington to Haig as she took him round Downing Street; President Galtieri's, to expose him to mass demonstrations.
17. In 1981 it had been assumed that this would be the case, but the base in South Georgia was reprieved. Argentina may not have been aware of this.

18. Roger Perkins, *Operation Paraquet: The Battle for South Georgia* (Chippenham: Picton Publishing, 1986).
19. House of Commons, *The Falklands Campaign: A Digest of Debates in the House of Commons 2 April to 15 June 1982,* (London: HMSO, 1982).
20. Cmnd 8787, *Falkland Islands Review* (London: HMSO,1983), para. 296.
21. Virginia Gamba, *The Falklands/Malvinas War: A Model for North-South Crisis Prevention* (London: Allen and Unwin, 1987).
22. Third Report of the Foreign Affairs committee, Session 1984–5, *Events of the Weekend of 1st and 2nd May 1982* (London: HMSO, 1985), p. 96.
23. Major General Sir Jeremy Moore and Admiral Sir John Woodward, 'The Falklands Experience', *Journal of the Royal United Services Institute* 128 (1983).
24. Peter Greig, 'Revelations', *Granta* 15 (1985).
25. Sir Nicholas Henderson, 'America and the Falklands', *Economist* 12 November 1983.
26. See Desmond Rice and Arthur Gavshon, *The Sinking of the Belgrano* (London: Secker and Warburg, 1984).
27. Article 51 states that:

> Nothing in the present Charter shall impair the inherent right of individual or collective self-defence if an armed attack occurs against a member of the United Nations, until the Security Council has taken the measures necessary to maintain international peace and security.

In Resolution 502 the Security Council expressed itself to be disturbed by reports of an invasion by the armed forces of Argentina, and determined that there had been a 'breach of the peace in the region' before demanding (1) 'an immediate cessation of hostilities' and (2) 'an immediate withdrawal of all Argentine forces from the Falkland Islands (Islas Malvinas)'. It concluded with a call to both sides to seek a diplomatic settlement. As Argentine forces did not withdraw, Britain felt able to exercise its 'inherent right of self-defence'.

9 The Falklands War and British Defence Policy

Paul Rogers

Britain ended the Second World War as a victor but with a severely crippled economy. After demobilisation, attempts were made in the late 1940s to bring defence spending down to a level which the economy could bear, but this ended with the rearmament programme ordered by the Attlee government in response to the development of the Cold War and, especially, the outbreak of the conflict in Korea. In the early 1950s defence spending remained close to the Korean War peak, but much of this was directed towards developing and procuring the weapons systems considered necessary to maintain Britain in the forefront of the world's military powers. A common perception among a majority of the population was that there were three, not two, superpowers, and much of the commitment to military spending was related to this perception.

This was especially true of the desire to develop nuclear and thermonuclear weapons as well as the delivery systems. In the mid-1950s, the large nuclear weapons development programme was therefore backed up by the simultaneous development of three different types of medium-range subsonic bombers, the so-called V-bombers, but Britain also committed huge scientific and technological resources to other areas of military R & D, especially interceptors, air defence missiles and armoured vehicles. Accompanying this high-technology commitment was the need to maintain the security of the Empire, a policy which required continued conscription and the maintenance of garrisons in many countries. This, combined with a perceived need to maintain great power status, was responsible for a major strategic commitment to the Middle East and South East Asia, including a force of some 80,000 troops in the Suez Canal zone.

The Suez crisis of 1956 was the most important single factor in curbing Britain's global security ambitions, and the development of defence policy in the following year, under Duncan Sandys, involved a much greater reliance on nuclear weapons as a means of maintaining status and security. This was accompanied by the end of conscription and the development of a professional army both involving quite marked cuts in overall defence budgets. In practice, though, defence budgets crept up again in the early 1960s as new roles developed and advanced military technology became more costly.[1]

A decade later, the Labour government under Harold Wilson regarded it as a major priority to bring Britain's defence spending down to a level which it believed the country could afford, and this involved an almost total re-orientation to Europe, the abandonment of a significant strategic role East of Suez and progressive cuts in defence spending. While the Heath government of the early 1970s came to power with the stated aim of restoring some of the roles trimmed by Labour, in practice most were left to one side. Britain thus concentrated on the European theatre, with most of its defence posture, and indeed equipment, related to the possibility of a conflict with the Warsaw Pact on the Central Front. This persisted under Labour until the end of the decade.

Even though the global role had been largely abandoned, there were still some significant exceptions, partly resulting from the legacy of the East of Suez posture and partly as a result of post-colonial anomalies. Thus, there remained a small garrison in Hong Kong, a small naval presence in the Persian Gulf and a large base in Cyprus. Political circumstances in the Caribbean involved maintenance of a frigate and a small army and air force presence in Belize; Gibraltar was a useful base with a continuing strategic significance; Ascension Island was used primarily by the United States for intelligence purposes; and a small Marines detachment was maintained in the Falkland Islands.

The remnants of the global role had two interesting aspects which were to become significant during the Falklands War. One was the location of important intelligence-gathering assets in former or current colonies. Facilities in Cyprus and Hong Kong were particularly useful, and Britain's Government Communications Headquarters (GCHQ) was a source of much valuable signals intelligence for the US National Security Agency.[2] This work enabled British operatives to retain close working links with their US counterparts, making it more easy to 'call in some favours' after the outbreak of the Falklands conflict. The second was a legacy of Britain's previous global naval role – the retention of a naval air arm through the building of so-called 'through-deck cruisers', in reality small aircraft carriers which could carry a small number of a navalised version of the Harrier STOVL aircraft. Without this naval air arm, the subject of some controversy during the 1970s, Britain would not have been able to undertake the Falklands campaign.

By 1979, tensions between NATO and the Warsaw Pact were increasing, and one of the last acts of the Callaghan government was to accept a NATO proposal that member states increase their defence budgets by 3 per cent per year. During the general election campaign, the Conservatives laid some emphasis on increasing Britain's defence capabilities but, in the event, their major action after the election was to grant a substantial pay increase, of around 33 per cent to the armed forces. As a consequence, and as economic pressures mounted over the period 1979–81, there was relatively little money available for increases in particular commitments. By mid–1980, the focus of the defence debate was almost entirely on East-West relations, with controversy developing over the decision to deploy mobile cruise and Pershing 2 missiles in Europe under US control.

The government announced in July 1980 that the UK would purchase the Trident C4 submarine-launched ballistic missile system to replace the existing Polaris force, but it became apparent within months that the production of the C4 system would not continue long enough for British purposes. In March of the following year, the government therefore announced the decision to purchase the more advanced, and more expensive, Trident D5 missile system, claimed by many anti-nuclear campaigners to be a 'first strike' weapon because of its accuracy. During 1981, the nuclear issue became a central feature of British politics, with large demonstrations against cruise missiles at Greenham Common and against the Trident programme, leading to the Labour Party adopting a unilateralist policy on nuclear weapons. This led to bitter differences within the party and was instrumental in the breakaway of the group who formed the SDP. Over the same period, though, the Thatcher government entered a period of deep unpopularity, primarily as a result of the worsening economic situation.

Although defence spending had been increased substantially in the first two years of the Thatcher government, it also made efforts to control spending by increasing efficiency. This became more urgent as economic problems for the government grew rapidly in 1981, and a new Defence Secretary, John Nott, was appointed with strong monetarist views. Spending on the nuclear deterrent was sacrosanct, so Nott commenced a review of conventional forces. The Royal Air Force's massive Tornado re-equipment programme was already far advanced and was, in any case, central to Britain's commitment to NATO on the Central Front. The same was true of the British Army, especially in the form of BAOR in West Germany, but with a continuing and costly commitment in Northern Ireland. Consequently, the main element of Nott's 1981 defence review was a re-assessment of the role of the Royal Navy.

At the time, the navy had three main roles: maintaining the Polaris fleet; a substantial anti-submarine warfare capacity designed primarily for action against Soviet submarines threatening transatlantic supply convoys in time of war; and the provision of an amphibious capability. Central to the full maintenance of the latter two roles were the aircraft carriers. In 1981 these comprised the elderly *Hermes* and the new *Invincible*, the first of the 'through-deck cruisers' planned and developed in the early 1970s. Two more *Invincible*-class carriers were being built and when these were completed, *Hermes* would be sold. As a result of Nott's review, it was agreed in 1981 that the navy's anti-submarine role would be cut back considerably. The number of escorts (destroyers and frigates) would be cut from sixty to forty-two, although the submarine programme would be largely maintained, *Hermes* would be scrapped and *Invincible* sold.[3] The future of the second and third of the new carriers, *Illustrious* and *Ark Royal*, did not seem bright. The future of the navy's amphibious sea-lift capability was not fully addressed as a decision to replace the two large amphibious warfare ships, *Fearless* and *Intrepid*, was not yet due.

The Nott review generated considerable controversy, not least because of the traditionally strong navy lobby among Conservative members in the House of Commons, and led to the resignation of the Navy Minister, Keith Speed. It was particularly unwelcome to many Conservatives because it effectively signalled the end of Britain maintaining any kind of significant global naval presence.

The Argentine decision to invade the Falkland Islands caught the British military authorities largely by surprise, but a decision was taken within twenty-four hours to assemble the forces necessary to re-take the islands. An impressive array of naval forces was assembled and set sail for the South Atlantic within four days, but these were, in reality, only the forward elements, a carrier task group of two carriers and escorts which would be required to offer naval protection and air cover to a much more substantial amphibious Task Force which was assembled over the following two weeks.

In retrospect, one of the key elements which allowed the Thatcher government strong national support in this risky endeavour was the decision of the Labour leader, Michael Foot, to offer unqualified support to the government during a brief parliamentary session held on the Saturday immediately after the Argentine invasion, a decision which led to bitter divisions within his party. In the absence abroad of the Shadow Foreign Secretary, Denis Healey, who would probably have counselled more caution, this bi-partisan support was invaluable in enabling the government to mobilise the necessary political and economic commitment to ensure that the massive task of assembling a substantial Task Force was completed rapidly.

The carrier force arrived in the South Atlantic by the end of April, hostilities commenced on 1 May, the amphibious Task Force was ready to land on the islands some three weeks later and a short, intense campaign followed which resulted in the re-taking of the islands. As this chapter is concerned with the broad issues of the effects of the war on British defence policy, the aim here will not be to examine all the complex details of the campaign itself, but rather to draw out some of the short- and long-term effects of the war.

The war itself involved a commitment which was only just within Britain's military capabilities at the time, bearing in mind that the onset of the South Atlantic winter was imminent and likely to add considerably to the logistical problems of re-taking the islands. There were also some significant immediate lessons from the conflict. The campaign involved the need to maintain air superiority over the islands, initially using carrier-based Harrier aircraft which also had to fulfil the role of ground attack aircraft. While the Argentinian air force had to operate at close to its maximum range, the Task Force had to operate its aircraft from the carriers stationed well to the east of the islands, out of range of Argentinian aircraft. On occasions, air cover was inadequate, especially at Goose Green, Bluff Cove and San Carlos, but the performance of the Harriers appears to have exceeded expectations, especially in relation to the success in keeping planes operationally available.

At the same time, the air cover available from the carriers was not adequate to protect ships or troops, and it also became clear that the point defence systems available to the Royal Navy were inadequate. Serious losses were incurred, and these would have been very much higher if Argentine munitions had been correctly fused. An immediate effect of the naval losses was the arming of Royal Navy escorts with light calibre anti-aircraft weapons. Indeed, in the campaign itself, such systems were hastily improvised. In the longer term, more attention was paid to close in-board weapons systems (CIWS) such as Phalanx and Goalkeeper, for the navy's aircraft carriers and escorts. These systems also offered some protection against anti-ship sea-skimming missiles such as the Exocet which destroyed *Sheffield* and *Atlantic Conveyor* and damaged *Glamorgan*.

It may be argued that the degree of protection afforded to British ships in the Persian Gulf from the late 1980s onwards was partly a result of the Falklands experience. The navy certainly made every effort to ensure that the ships sent into such a high-threat environment were those which had been retrofitted with a CIWS.

A second area of experience concerns ground operations in difficult environments. The offensive against the main Argentine forces near Port Stanley had to be mounted, at least in part, on foot, as air-lift capability was painfully inadequate. This partly stemmed from the loss of the heavy-lift helicopters on the *Atlantic Conveyor*, supporting the contention that the loss of that ship was actually the most important Argentine military action of the entire campaign.

Although not widely discussed then or since, the other major British problem was the failure of the nuclear-powered hunter-killer submarines (SSN) to provide an effective forward shield for the task force. The quiet Argentine diesel-powered boats, especially the Type 209, caused considerable concern to British ASW specialists, and it was intended to provide a forward shield of up to five SSN between the task force and Argentine naval bases. In the event, a shortage of SSN meant that it was not possible to deploy more than three for most of the war.[4] Furthermore, of the three SSN deployed, one experienced a mechanical fault which resulted in its being inoperable, and a second operated on half power for most of the period. Thus, the forward SSN shield was limited in extent. In the event, Argentina too had major problems with its submarines and the actual threat was therefore relatively small.

During and after the war, one of the key logistical problems was the distance involved in the air bridge to the South Atlantic, and this placed heavy strains on air transport and air-to-air refuelling. As a result, the RAF instituted a major programme of tanker upgrading after the conflict.

Another problem was that the rate of use of munitions throughout the conflict was, reportedly, well above the expectations of planners, and this, too, resulted in a post-war reassessment of logistical support for armed forces in times of conflict. It appears, though, that it proved difficult to invest in additional munitions stocks. Experience with the UK build-up of forces in Saudi Arabia in the autumn of 1990 suggested that stocks available would not be sufficient for training and combat purposes in that period, and supplies were sought from NATO allies not involved in the deployment of ground forces to the area.

As Lawrence Freedman argues, one of the most interesting aspects of the Falklands conflict was the possible role of nuclear weapons. At the time of the conflict, the Royal Navy had two types of nuclear weapon in service, the tactical free-fall WE177 bomb and the Polaris submarine-launched ballistic missile. The WE177 came in two variants, a version for surface attack against maritime or land targets delivered by Sea Harrier strike aircraft, and an anti-submarine depth-bomb version for delivery by ASW helicopters. The Polaris SLBM was deployed on four *Resolution*-class SSBN, whereas the WE177 could be deployed on aircraft carriers and on frigates and destroyers.[5] Some escorts previously involved in a NATO exercise in the Western Mediterranean sailed for the South Atlantic in early April 1982, carrying the ASW version of the WE177 nuclear bomb, as did some ships from Britain. Indeed, one frigate captain refused to sail from a British port without his ship's complement of tactical nuclear weapons.[6]

There appears to have been considerable controversy within

the Ministry of Defence over this matter, partly because the Royal Navy had very few tactical nuclear weapons and risked losing some of these if ships were sunk in the war zone, but there may also have been concern over the consequences of using tactical nuclear weapons against a non-nuclear power. In the event, there are indications that most, and possibly all, the tactical nuclear weapons were removed from ships at Ascension Island and transferred to a Royal Fleet Auxiliary munitions ship, *RFA Regent*, which, while it then deployed to the South Atlantic, was kept out of the immediate war zone. After the Falklands conflict, just two classes of Royal Navy surface ship were authorised to carry tactical nuclear weapons in peacetime, the *Invincible*-class carriers and the Type-22 frigates. This may have been a result of the internal controversy during the conflict.

Concerning strategic nuclear weapons, there is evidence, but largely in the form of 'leaks' from parliamentary, military and civil service sources, that a Polaris SSBN was deployed south of the Equator during the height of the conflict. In the absence of conclusive documentary evidence to this effect, though, it remains more a matter for speculation.

Finally, the war itself left a yawning gap in Britain's own defence posture. The term 'Task Force' is, in reality, a misnomer, for it was a force substantially larger than any Western European navy apart from that of France. It involved both of Britain's aircraft carriers, both assault ships and the great majority of the modern destroyers and frigates. The latter included five out of eight Type 42 destroyers, six out of eight Type 21 frigates and both Type 22 frigates then in service. Five out of eight operational SSN were eventually deployed as was almost the entire Royal Fleet Auxiliary. Britain's only commando brigade was deployed as was the army's rapid deployment 5 Brigade. The RAF's tanker fleet was put under immense pressure as was a substantial proportion of Britain's merchant marine. In essence, the operation left Britain largely without a navy for several months.

The defence commitment to the Falklands continued at a lower level for several years. Six months after the war, the Falklands garrison stood at 7,000. Even three years after it still involved 4,000, with hundreds more at Ascension Island. For some two years, the Falklands naval patrol involved five frigates and destroyers at any one time, normally the most modern ships in the navy. Allowing for time of passage, replenishment and repairs, this actually involved a permanent commitment of some thirteen ships, representing at least 25 per cent of the navy's entire force of escorts.[7] Developing and maintaining the Falklands garrison involved the provision of large numbers of specialists, particularly in the first two years. Some seven squadrons of Royal Engineers, 1,400 men in all, were involved in building and installing facilities and equipment, severely limiting British Army capabilities in West Germany. Maintaining the garrison required facilities for many aircraft types including Phantoms, Hercules, Harriers, Gazelles, Chinooks, Sea Kings, Wessex and Lynx, involving a wide range of spares and specialist equipment. Even in 1991, with a smaller range of aircraft, some six types had to be serviced and repaired.

The decision to build a major new air base at Mount Pleasant involved immediate financial costs although it ultimately resulted in the maintenance of a smaller garrison. The cost breakdown for the war and the longer-term Falklands defence commitments is, according to official sources, as shown in Table 9.1. This suggests a figure of about 3.4 billion pounds over the six years.[8]

TABLE 9.1 *Costs of the Falkland's defence commitments*

1982–83	800 million pounds
1983–84	624
1984–85	684
1985–86	552
1986–87	450
1987–88	300

The effect of the Falklands War on British defence policy in general can be considered in relation to three themes. The first was the particular lessons learnt, such as the need for improved naval point defence, improved air tanker support and the requirements for more heavy-lift helicopters. As already mentioned, improved naval point defence became relevant to operations in the Persian Gulf, but this also applied to air tanker support. In the immediate post-Falklands War period, the prevailing view would ordinarily have been that there was no need to improve the RAF's aerial refuelling capability, as this was not justified by operations conducted within NATO. In practice, though, the need to re-supply the Falklands prior to the completion of the Mount Pleasant air base was such as to require a further investment in air tanker support. As a result, the RAF was particularly well-served by air tanker support by the end of the 1980s, indeed more so than European operations required. This meant that, at the onset of the Gulf crisis in the autumn of 1990, the RAF was able to deploy substantial tanker forces to the Gulf, and these were used extensively after the war started early in 1991.

The second was the role of the navy in the successful completion of the campaign, since this effectively ended plans to cut the carrier and escort force. The ships lost in the war were replaced with newer escorts, and the *Invincible*-class of three carriers was completed and deployed with the Royal Navy. General Galtieri effectively spiked John Nott's plans for the navy.

Once again, though, there was an interesting aspect of this apparent early in the Gulf crisis in 1990. Britain's early commitment of naval forces to the Gulf included a number of escorts, minehunters and auxiliaries, but did not include an aircraft carrier. Royal Navy lobbyists were anxious to commit a carrier to the Middle East force but were apparently overruled, and the major British commitment was concentrated on a reinforced armoured division and a substantial force of RAF strike aircraft and interceptors. Thus, unlike the Falklands, the Royal Navy was the junior partner in the Gulf operations, a situation which was not to the liking of the sea lords. Indeed, shortly before the war itself started, *Ark Royal* even went as far as conducting exercises in the Eastern Mediterranean, an apprent effort by the Royal Navy to be 'on hand' in the event that such a ship was required.

Finally, and in more general terms, the success of the British action in the Falklands War resulted in more stress being put on global reach or, to use the more anodyne term, 'out of area' operations. To the Thatcher government, the Falklands success showed that Britain was still a significant military power and this tended to put off, perhaps by five or more years, a reconsideration of the longer-term needs of British security.

In political terms, the perceived success of out of area operations in the early 1980s was almost certainly influential in determining Britain's response to the Iraqi invasion of Kuwait in August 1990. Britain had maintained a significant naval presence in the Gulf for many years, but was quick to support the United States in its build-up of forces in the area, agreeing first to send interceptor and strike aircraft and then an armoured brigade. By late September 1990, a major operation was underway to air-lift and ship a substantial military force to Saudi Arabia and the surrounding states, with Britain deploying the largest military force of any West European ally of the United States. In early November, the Bush administration decided to add massively to the US forces deployed to the region and Margaret Thatcher, in one of her last decisions before her resignation in November 1990, ordered a doubling of British ground forces in the Gulf to bring them to the level of a reinforced armoured division. On the outbreak of hostilities in the Gulf on 16 January 1991, Britain had some 40,000 military personnel in the Middle East, by far the largest number since the mid-1950s.

Throughout a fairly wide spectrum of British politics, the deployments to the Gulf seemed to represent proof that Britain was still a major military force in the world, and the relatively lower commitment of many European NATO allies was the cause of much criticism. Even so, wider realities were already at work. In mid-1990, with the easing of Cold War tensions, the Ministry of Defence had already embarked on a modest defence review, though this was less rigorous than those being undertaken in many allied countries. It appeared that British defence spending in the 1990s, while likely to decrease, would still remain well above the average for its NATO allies. In some small measure, the Falklands War played a part in this. Perhaps it presented a useful excuse for failing to face up to the reality of excessive defence spending continuing to handicap Britain's industrial economy. Maintaining a global role was intrinsically attractive to the British political culture.

It was therefore ironic that, no sooner was the Gulf War over, then the government's long-awaited defence review announced the smallest British Army since the 1830s, together with sharp cuts in both the other services. In the final analysis it may well be that global developments, notably the decline of the Soviet Union, rather than regional conflict in the Middle East, will be the true determinant of Britain's defence policy in the mid and late 1990s. If so, the Falklands War will have done no more than give a false promise of a continued global status which was, in reality, already a matter of history.

NOTES

1. For an indication of defence spending over the period 1948 to 1979, including constant price figures, see Appendix IV of Michael Dockrill, *British Defence Since 1945* (Oxford: Basil Blackwell, 1988).
2. A comprehensive analysis of UK/US intelligence relationships may be found in Jeffrey T. Richelson and Desmond Ball, *The Ties That Bind* (London: Allen and Unwin, 1985).
3. See Dockrill, p. 115.
4. Details of UK SSN operations in the conflict were contained in a diary kept by an officer on *HMS Conqueror*, Lt. Narendra Sethia, substantial sections of which were published in a number of US newspapers in 1984, although not with the approval of the author.
5. See Paul Rogers, *Guide to Nuclear Weapons*, (Oxford: Berg, 1988), pp. 65–6 and 70–2.
6. A more detailed discussion of this issue is found in the Belgrano Action Group's *The Unnecessary War*, (Nottingham: Spokesman Books, 1988), especially pp. 96–108.
7. See Paul Rogers, *A Note on Future Argentine Military Postures in Relation to the Falkland Islands*, in Third Report from the Defence Committee of the House of Commons, 1982–3, on *The Future Defence of the Falkland Islands* (London: HMSO,1983), pp. 202–5.
8. See Third Report from the Foreign Affairs Committee of the House of Commons, Session 1984–5, on *Events Surrounding the Weekend of 1–2 May 1982* (London: HMSO, 1985), pp. cxiv–cxv.

10 Anglo-American Relations and the Falklands Conflict

Christoph Bluth

It has been one of the peculiar features of the 'special relationship' between the United States and the United Kingdom that however one judges its success in co-operation for the defence of Europe, its extension to out-of-area issues has often been difficult as American and British interests in various regions of the world have often differed. One historical parallel to the Falklands conflict of 1982 which has been cited frequently is the Suez crisis of 1956 where the United States strongly opposed British policy. This has been interpreted in different ways. Some in the US administration, like Defence Secretary Caspar Weinberger and Secretary of State Alexander Haig saw Suez as an example not to be repeated. In their view the American stance during the Suez crisis had dramatically weakened Western unity after World War II and damaged the Anglo-American relationship. The Argentine junta on the other hand interpreted the precedent of Suez as evidence that the United States would be on the side of anti-colonialism and not support Britain. Although the United States eventually supported Britain very strongly over the Falklands, initial American responses were marked by the some degree of ambivalence as the administration found itself in a situation where it had to decide between competing strategic interests.

The United States had first become involved with the Falklands question in 1831 when Captain Duncan of the *USS Lexington* sacked the settlement of the governor of the islands appointed by Buenos Aires, Louis Vernet, because Vernet had dared to enforce fishing regulations against American vessels. Captain Duncan then declared the islands free of all government. The British takeover in 1833 took place with American knowledge and approval.[1]

Strangely, the United States did not object to the British

action on the basis of the Monroe doctrine, which prohibited the acquisition of new colonies by European powers in the Western hemisphere. However, Washington did not recognise British sovereignty over the Falklands. When Buenos Aires initiated an action for damages against the US government because of the actions by Captain Duncan, the Americans responded that the issue of damages could not be considered until the question of sovereignty was decided. Non-recognition thus served as a convenient excuse for avoiding or postponing indefinitely payment of damages, although an American court (in a separate action) had already established that Duncan's operation had been illegal.[2] Furthermore, it was convenient for the US not to recognise British sovereignty because Britain also attempted to enforce its fishing regulations in the area. In 1854, an American vessel was seized by the British in the territorial waters of the Falklands and subsequently released. The State Department protest against the seizure contained intimations that it did not consider the issue of sovereignty to be settled. This convenient device could be used against either Britain or Argentina whenever required. A State Department paper prepared at the time reaffirmed the neutral position of the United States with regard to the Falklands dispute. Even when the United States supported British military action in response to the Argentine invasion of the islands, this did not amount to a support of the British position with regard to the Falklands issue in general.

The dilemma which faced the Reagan administration when the Falklands dispute came to a head in 1982 was that both protagonists were considered allies of the United States. Argentina's relations with the United States improved considerably when Ronald Reagan assumed the presidency in the US and Roberto Eduardo Viola in Argentina (the latter in March 1981). Under Reagan, the United States undertook a major effort to restore its strategic position in the world which was thought to have been significantly eroded during the 1970s. This involved a substantial build-up in military forces, both with regard to strategic nuclear forces and conventional intervention capabilities. It was accompanied by an attempt to overcome the 'Vietnam syndrome' and relegitimise military intervention in the context of a strategy of a new effort to contain the expansion of Soviet power. This implied advocating the legitimacy of and developing the capability for a wide range of military, political and economic actions against radical movements and states in the Third World. This policy later became known as the Reagan Doctrine.[3] The Carter human rights policy was abandoned to be replaced by a policy of open support for right-wing regimes in Latin America, in the hope of cementing an anti-Communist alliance in the Western hemisphere. The Humphrey-Kennedy sanctions imposed on Argentina during the Carter administration because of the appalling human rights record of the military regime were lifted. Viola visited the US shortly before he assumed power on invitation from President Reagan. General Galtieri visited the US in August 1981 at the invitation of the US Army Commander-in-Chief, General Meyer, and in October for the inter-American Conference of Army Commanders-in-Chief. During his second visit he was treated almost like an important head of state, having dinner with Caspar Weinberger, and meetings with national security adviser Richard Allen and the American ambassador at the United Nations. Galtieri was clearly given the impression that the relationship with Argentina had assumed great importance for the US administration. This was linked to the deteriorating situation in Central America. The American administration felt that its hands were being tied by Congress which had limited the number of American 'advisers' permitted in El Salvador to fifty-seven. General Galtieri enthusiastically took up this opportunity for direct co-operation with the United States and sent some of his troops. About 500 Argentinian soldiers participated in sabotage raids in Nicaragua, operating mainly out of Honduras. Furthermore, Argentina was seen as an ally ensuring security in the South Atlantic against Soviet maritime power. The Argentine government promoted the concept of a South Atlantic Treaty Organisation involving South Africa and the United States.[4]

The most significant aspect of these visits from the perspective of the Falklands/Malvinas conflict lies in the fact that they were used by the Argentines to 'test the diplomatic waters' and see how the United States would react to an Argentine operation to recover the Malvinas. They became convinced that the United States would look the other way and not object in return for co-operation in Central America and the South Atlantic.

It is widely believed in Latin American circles that there was a more explicit understanding. Thus the Labour MP Tam Dalyell has claimed in his book *One Man's Falklands* that General Vernon Walters, who made a visit to Argentina in 1981 as President Reagan's emissary, discussed the setting up of a South Atlantic Treaty Organisation, and that there was even talk about setting up a military base on the Falklands similar to the American naval base on the island of Diego Garcia in the Indian Ocean.

> However, the understanding was that the Agreement on hemispheric and other grounds, should be between the United States and Argentina, the bulwark of American policy in the South Atlantic, and not between the United States and Britain. Asked by the Argentine military what Britain would do, the Americans replied to the effect that the British would 'huff, and puff, and protest and do nothing', with the implication that the Americans could soothe ruffled British feathers.[5]

Similar speculations appeared in the *Latin American Weekly Report* in March 1982:

> [Argentine] Government officials feel that the international repercussions of a hard line against Britain (i.e. invasion) will be manageable . . . in the light of Washington's preoccupation with security in the South Atlantic . . . [Washington] . . . would be happy to see the issue settled. It could then open the way to the installation of US military bases in the South Atlantic . . . a development over which there has been much speculation in Buenos Aires since Galtieri took power . . . [6]

This article was based on a press release by the Argentine government and it is clear therefore that in early March 1982 Argentine government officials were publicly talking about an American base on the Malvinas.

No evidence has emerged from the American side that there was any kind of understanding of this nature between the United States government and the government of Argentina. The idea of a South Atlantic Treaty Organisation was promoted actively by Argentina, but found an unenthusiastic response in Britain and an indecisive one in the US. While Argentina clearly overestimated its importance to the United States, and, more seriously, misinterpreted the lack of serious objections whenever subtle hints about the Malvinas question were made as American approval, it is nevertheless true that relations with Argentina had become much more important to the US. This was particularly the case because it proved to be difficult to generate support for US policy in Central America from other Latin American states. Chile was in conflict with Argentina over the Beagle Channel and relations with Chile were therefore difficult. Brazil was unwilling to support US policy because it was incompatible with the relations it was seeking to develop with Third World countries. Venezuela's radical President was seeking to form a new Latin American grouping in co-operation with Cuba which would exclude the United States. Relations with Mexico were strained because of the way in which they had been mishandled under the Carter administration (in particular with relation to the gas pipeline project and the immigration issue).[7] Nevertheless, one of the principal miscalculations by the Galtieri regime when it decided to invade the islands was with regard to the American response.

That response was at first ambiguous. The ambiguity arose out of the importance of both Britain and Argentina for US foreign policy. Britain's importance as one of America's most loyal NATO allies really goes without saying. It has to be remembered that the Falklands conflict took place at a time when the principal European allies of the United States were facing a deep internal controversy over the future of nuclear defence policy and transatlantic relations. As the Falklands crisis broke, the CIA and perhaps the State Department feared that the Thatcher government might fall and an anti-nuclear Labour government take its place.

At first, shortly before the Argentine invasion, the US adopted an 'even-handed' approach, arguing for restraint on both sides, a stance which caused much anger in the Foreign and Commonwealth Office. Mrs Thatcher then managed to enlist President Reagan's help for a last-minute appeal to Galtieri not to resort to force – unsuccessfully. Apart from this particular event, the support Britain expected from the US was not forthcoming. British officials were particularly angered that on 2 April, the night of the Argentine invasion, top American officials including Jeane Kirkpatrick, Walter Stoessel (Deputy Secretary of State) and Thomas Enders (Assistant Secretary of State for Latin American Affairs) attended a banquet given by the Argentine Ambassador Esteban Takacs. Kirkpatrick, a specialist on Argentine politics, did nothing to improve things in the eyes of the British by commenting in an interview a few days later, 'If the Argentines owned the Falklands, the moving in of troops is not armed aggression.'[8]

The Falklands crisis sparked off a fierce dispute among American foreign policy-makers. Thomas Enders argued that the United States should not jeopardise its favourable relations with Latin American states just because of the Falklands dispute, bearing in mind the problems in Central America. Lawrence Eagleburger, Assistant Secretary of State for European Affairs, put the case in favour of supporting a NATO ally. American ambivalence reflected not only the different interests of different departments in the government bureaucracy, but it was also inherent in US commitments under the Rio Treaty (the Inter-American Treaty of Reciprocal Aid), a mutual defence pact between the United States and Latin American countries, and the NATO treaty involving Britain. These commitments, which had always been part of the same global strategy, suddenly appeared to be in conflict. Alexander Haig, the Secretary of State, despite his strong support for American foreign policy in Latin America (of which he was one of the architects) believed that ultimately the US should support Britain, but he also saw the possibility of American mediation between two allies, with himself in the role of mediator.[9] At a meeting of the National Security Council on 7 April Haig's suggestion to attempt a mediation in the dispute was approved by the President.[10]

While Haig and his team were on their way to London on 8 April, Defence Secretary John Nott announced a maritime exclusion zone around the Falkland Islands. This was interpreted as a deliberate signal that Britain was adopting a firm attitude. Indeed, the War Cabinet went to great lengths to impress Haig with the firmness of its stance. There could be no deviation from the objective that Argentina should honour the UN Security Council Resolution 502. Until such time Britain would sustain her rights under Article 51 of the UN Charter. Haig seemingly did not need much persuasion. In his memoirs, he emphasises his admiration for the British and their stance against aggression. He confides that he did not come to London to urge a compromise of principle on Mrs Thatcher. He recognised the supreme importance of democracies demonstrating their strength and assured Mrs Thatcher that there would be no repetition of Suez. Indeed, he was in London 'to help the British' and assured Mrs Thatcher of his full sensitivity 'to the depth of British feelings' on the Falklands.[11] However, he argued that he needed a bit more room for negotiations. He presented the three central ideas of his team as broad goals for the negotiating effort:

(1) Mutual withdrawal of forces (i.e. Argentine troops would be withdrawn from the Islands and the Task Force would not be deployed close to the Islands).
(2) Some interim authority for the Islands.
(3) Negotiations to determine the status of the Islands by a certain date.

Haig 'pressed on the question of interposing an international force on the islands and setting up some sort of interim administration',[12] but Margaret Thatcher was sceptical. She thought that the American ideas were 'just too woolly'.[13] Haig understood the political background which constrained Thatcher's approach to the negotiations. From the day of the Argentine invasion, she had been under tremendous political pressure to redeem herself after the failure of her Falklands policy and her inability to foresee and prevent the Argentine action. On 3 April she managed to stave off parliamentary wrath by promising to redress the situation. Specifically, she promised that British territory would be recovered and that the wishes of the islanders would be paramount in any settlement. Furthermore, the government had based its public justification for the sending of the task force on the principle that aggression must not pay and Argentina must not be seen to gain anything by it.[14] Those three principles would form the basis of Parliament's judgement as to her success. They left precious little room for manoeuvre in negotiations. The insistence that the wishes of the islanders must be paramount implied a return to the pre-invasion *status quo* (British rule). In any case, the British government was not willing to discuss the future of the Falklands with Argentina before the withdrawal of Argentine troops. Britain was not willing to negotiate under duress.

In his report to President Reagan, Haig indicated that some shift in the British attitude was necessary for the negotiations to succeed and bloodshed to be avoided.[15] He understood that the junta could not survive a complete loss of face and needed something tangible in return for a peaceful withdrawal. Before he came to London, the British Ambassador to the United States, Sir Nicholas Henderson, who had begun to engage in a sustained and successful public relations effort to persuade the American public of the merits of Britain's case, had told him in no uncertain terms that the British government had no interest in Galtieri's political survival.[16] Haig, however, knew that a negotiated settlement could only be achieved if both governments could live with it.

When the 'shuttle team' arrived in Buenos Aires, the Plaza de Mayo was packed with vast demonstrations in support of the Argentine action to 'recover' the Malvinas. In his talks with Galtieri, Haig tried very hard to convince him that the British were serious about military action to retake the islands, and that in the final analysis the United States would have to back Britain. Despite his shock about the swiftness and strength of the British response, and the inaccuracy of all the predictions with respect to the diplomatic responses to the Argentine action, Galtieri would not believe that Britain's warnings were real. Argentina's claim to sovereignty over the Malvinas could not be conceded; the junta was reliably informed that Britain would not retaliate. Admiral Anaya believed the British fleet would 'break down' and could not mount a successful operation to retake the Islands. The basic position of the junta was that Britain should lift sanctions, the Task Force should be stopped and a specific date for the conclusion of negotiations about sovereignty should be set.

The general outline of the kind of arrangement which Haig envisaged might remove the threat of armed conflict assumed that the issue of sovereignty was to be deferred. There should be a simultaneous withdrawal of forces supervised by an international consortium. This would involve the creation of a zone from which British forces would be excluded. There would be an international interim administration, while the settlement of the dispute would be negotiated.

The principal stumbling-block during the first round of the Haig mediation effort was that the junta insisted that some residue of Argentine authority must remain on the islands, such as an Argentine governor. Any progress which

Haig believed was made on these issues turned out to be illusory. Just before Haig departed for London, Foreign Minister Costa Mendez handed him a paper with 'some personal thoughts of his own'. They represented in fact the official Argentine position as formulated by the Military Committee.[17] The paper presented two alternative options which would be accepted by Argentina. The first one made no preconditions about sovereignty, but would involve an Argentine governor and the flying of the Argentine flag on the islands. The alternative option was a commitment to negotiations with a guarantee that they would result in the transfer of sovereignty to Argentina by 31 December 1982. During the interim period, demilitarisation must take place in accordance with Resolution 502 and Argentine citizens should have the same rights as British citizens to enter the islands.[18]

In London, the Argentine proposals were considered quite unacceptable. Even the various schemes drafted by the Haig team found little support. An 'interim period' for negotiations might be possible so long as a 'recognizably British' administration was restored to the islands with at most a very minor Argentine role. The Argentine representative could fly the Argentine flag but not all over the islands. Britain could not accept the withdrawal of her forces to operational bases 8,000 miles away. Any mutual withdrawal of forces (for example 400 miles – the distance to the Argentine bases) was problematic since the British fleet had no operational base and the British were not enamoured of the notion of having their fleet move around the South Atlantic indefinitely.

Meanwhile Haig continued to keep in contact with Costa Mendez. On 14 April Costa Mendez made a proposal which to Haig appeared to contain a small degree of movement. In Haig's words:

> His government was now willing to modify its demands. It would not insist on an Argentinian governor if the agreement contained a British acknowledgement that it intended to "decolonize" the Falklands in compliance with the 1964 United Nations Declaration on Decolonization. However, Buenos Aires was not prepared to submit this proposal formally without some advance indication of flexibility from London. Argentina also required guarantee that the British fleet would limit

> its movements and, from the United States, a firm statement that the U.S. was not assisting Britain militarily in any.[19]

Haig states in his memoirs that neither he nor Francis Pym was 'immediately familiar with the U.N. Declaration mentioned by Costa Mendez'.[20] This is a strange admission if we consider that the UN declarations and the Resolution are so fundamental to the whole history of the dispute that they are discussed in every Argentine propaganda leaflet. Haig goes on to state: 'I interpreted the Argentinian proposal to mean that Beunos Aires accepted autonomy and self-determination for the Falklanders.'[21] From Argentine accounts of the dispute it is clear that this is the precise opposite of what they meant.[22] Haig was so badly briefed about the Argentine position that he was unable to understand the language the Argentines were using.

The next round of the Haig shuttle therefore began on entirely false premises. Haig's strategy, as it unfolded during his second visit to Buenos Aires, consisted in seeking to persuade Argentina that its basic objectives could be satisfied by a formula which did not use language that promised a transfer of sovereignty to Argentina explicitly, but nevertheless would ultimately result in such an outcome.

One of the conceptual vehicles for Haig's approach was the principle of self-determination. Haig explained his view in the following manner:

> Britain was prepared, as she had been from the start, to negotiate a settlement as long as the islanders were given the opportunity of deciding the future. Once again, we had a reasonable alternative to useless bloodshed – and once again, its fate was in the hands of the Argentines.[23]

This obscured the fact that the issue of self-determination was *the* central issue of the dispute, and had been for over fifteen years. If self-determination for the islanders was accepted, and if Argentina could be prevented from flooding the islands with their own citizens, this would simply imply a reversion to British rule. Haig believed that self-determination was the one issue on which the British government would not concede, whereas Argentina might. As counter-intuitive as it may appear at first sight, and as much as official British statements might deny it, precisely the opposite was the case. The principle that the islanders did not have the right to self-determination was the one principle that Argentina was not willing to concede, whereas the British government was willing to fudge the issue.

It is clear that Britain was not upholding the islanders' right to self-determination before the conflict of 1982. She was actively engaged in negotiating a lease-back arrangement with Argentina whereby sovereignty would have formally been transferred to Argentina. This policy failed due to opposition in Parliament. Another action by the British government which cast doubt on the rights of the islanders to self-determination was the Nationality Act of 1981, which excluded the Falkland Islanders from full British citizenship.[24] Although after the Argentine invasion the principle of self-determination for the islanders was reasserted and therefore acquired much greater significance than hitherto, some flexibility on this issue can be detected in the course of the conflict. Thus Margaret Thatcher allowed a proposal to be presented to Argentina which would allow for an eventual transfer of sovereignty while safeguarding the islanders' right to determine their future.[25] It is interesting to observe how the concept of 'autonomy' and 'self-determination' came to be blurred. In the final stages of the Haig mediation, Haig used 'self-determination' to mean 'autonomy' or the islanders' freedom 'to determine their future'. Lawrence Freedman and Virginia Gamba-Stonehouse comment on Haig's approach to these conceptual issues: 'Central to this idea was the notion of self-determination, which Haig regarded as crucial to the British position and far more important that sovereignty itself'.[26] But this ignores the conceptual framework in which the Falklands/Malvinas controversy had been carried out prior to the 1982 conflict.

The crucial distinction lay between Argentina's position, to safeguard the 'interests' of the islanders, and the British position, which insisted that the islanders' 'wishes' were paramount. In this framework, self-determination implies sovereignty. There can be no self-determination for the islanders without British sovereignty. But the British were willing to fudge the issue after the end of the Haig meditation and the sinking of the *Sheffield*, when a new definition of self-determination was used by the FCO which no longer implied 'deciding who has sovereignty' but could be satisfied, as in Haig's proposals, by local self-government or some form of 'autonomy'. The point to observe here is that the distinction between 'autonomy' and 'self-determination' is similar to that between the 'interests' and the 'wishes' of the islanders. In the early stages of the Haig mediation the British government still insisted that ultimately the 'wishes' of the islanders had to be determinative in any settlement, which inevitably implied a return to British rule to the exclusion of Argentina. However, there was a shift in the British government's position during the conflict, while there was no shift in the Argentine position that self-determination cannot be accepted.

Haig went beyond what the British government could accept and gave Argentina strong hints that the negotiations would most likely resolve the entire issue in such a way as to achieve Argentina's principal objective, that is, the transfer of sovereignty. This was still not enough for the junta, which sought stronger guarantees for the outcome of the process. Confused decision-making in Buenos Aires increased the difficulties in the process where concessions were made and unmade without anyone appearing to have the definite authority to negotiate. But the failure of the negotiations had a deeper reason, as Freedman and Gamba-Stonehouse have pointed out:

> The very simplicity of the issue created its own difficulties, for it was not possible to draw on a range of interlocking issues to produce a series of complex trade-offs. Basic principles of sovereignty and self-determination were at stake and these do not leave a lot of room for compromise, especially with the addition of prestige.[27]

In addition, the junta was not convinced, even after Britain recovered the island of South Georgia on 25 April, that war would come or, if it did, that Britain would be successful. The political price of conceding the substance of Argentine demands appeared to be too high.

During the time of the Haig mediation the contradictions between American alliance commitments threatened to come out into the open when Argentina called for a meeting of the foreign ministers of the Organisation of American States (OAS) on 20 April under the Rio Treaty. The only OAS response to the Falklands/Malvinas crisis so far had been an innocuous resolution by its Permanent Council on 13 April offering co-operation to Argentina and the United Kingdom to reach a peaceful solution to the dispute. The meeting of the foreign ministers took place on 26 April (one day after the British assault on South Georgia). This was the first time that the Rio Treaty had been activated without involvement by the United States. Costa Mendez realised that he would not be able to get much in the way of active support from other Latin American countries. Hence he announced that Argentina was not at this stage requesting economic sanctions or military assistance against Britain. His goal was a resolution demanding the withdrawal of the Task Force.

The Resolution passed by the Consultation of Foreign Ministers of the OAS on 28 April was interpreted as a success for Argentina, although it did not go as far as the Argentines might have wished. Not only did it fail to demand the withdrawal of the British fleet (it merely called for a cessation of hostilities); it also mentioned Resolution 502 and implied that it must be complied with. The one feature of the Resolution which was clearly in Argentina's favour was that it recognised Argentine sovereignty over the Malvinas, as well as the need to uphold the islanders' interests (rather than wishes). The United States abstained from the vote as part of its ostensibly neutral approach during the mediation, after Haig had given a strong speech on the need to observe international law and the dangers ahead if a settlement was not reached soon, arguing that the resolution would not help the mediation process.[28]

After serious military conflict had broken out, Argentina managed to rally Latin American support to a greater extent, and on 28–9 May the OAS issued a Resolution condemning Britain and calling for a complete British withdrawal. The possible damage to American interests resulted in some efforts by the administration to begin a process of mending relations, which in the case of Argentina was rebuffed. The re-emerging disputes within the US Administration led to the remarkable scenes in the UN Security Council when Ambassador Jeane Kirkpatrick cast a veto against a ceasefire resolution and announced to the astonishment of the world that according to new instructions she would change the US vote to an abstention if a change of vote were possible.[29]

What is especially interesting from the point of view of this essay is the interplay of British and American interests in the mediation effort. One principal American interest was to prevent the outbreak of armed hostilities. Given his position as mediator, this was also very much in Haig's own personal interest. Furthermore, the relationship of the United States with many Latin American countries could be adversely affected by too close an association with a British war effort. At the same time, a British defeat was also against American interests. The supreme interest of the British government lay in a favourable resolution of the crisis. The survival of the Thatcher government depended upon it. There was very limited preparedness for concession, and a great deal of scepticism with regard to the mediation effort. Intriguingly, it is possible to interpret the mediation as working in Britain's favour, in as much as it allowed time for deployment in the area, and appeared to exhaust the alternatives to military action.

The end of the Haig mediation resulted in the official 'tilt' of the United States in favour of Britain. A corresponding resolution was put before the Senate on 29 April. The previous evening Haig had given a secret briefing on his mediation, in which he emphasised Argentine intransigence. During the debate on the resolution sponsored by Senators Joseph Biden and Daniel Patrick Moynihan, there was an attempt to dilute it, spearheaded by Senator Jesse Helms, backed by the Republican Senators John Warner and Paula Hawkins. A sentence according to which the US should 'further the efforts of the British Government to achieve in the Falkland Islands the full withdrawal of the Argentine forces' was changed to read 'to further all efforts pursuant to Security Council Resolution 502'.[30] The resolution was eventually carried by a majority of 79:1 (Helms voted against). Haig announced the failure of his mediation on 30 April and stated that the United States would apply military and economic sanctions against Argentina as well as respond positively to requests for material support from Britain.

We now know, however, that despite the official position of neutrality during the Haig mediation, the United States supplied substantial military aid to the British military effort from the very outset. Secretary of Defence Caspar Weinberger realised the tenuous nature of the logistics and supply line. Britain had no air surveillance and inadequate satellite communications. Intelligence capabilities left a great deal to be desired. For the engagements with the Argentine Air Force, Britain was short of surface-to-air missiles. In consequence, low-level, confidential co-operation between the British and American navies began even before the Task Force left for the South Atlantic. Weinberger decided from the outset that a British defeat could not be contemplated. In his own words:

> I therefore passed the word to the Department that all existing UK requests for military equipment, and other requests for equipment or other types of support, short of our actual participation in their military action, should be granted immediately.[31]

Direct American involvement in the crisis would have been politically very costly in terms of US interests in Latin America, but a decision was made that should the Argentines sink either of the British aircraft carriers, *Hermes* or *Invincible*, the United States would immediately provide a replacement.[32] Weinberger's decision to provide very substantial help to Britain was cleared by the President but apparently not by the full National Security Council.[33]

At first it was planned in London that the Task Force should move straight to the South Atlantic from Gibraltar, but it was soon realised that it would be necessary to use the facilities of the American Wideawake airbase on Ascension Island. Britain had leased the island to the United States and the leasing arrangement provided for the British right to use the base in case of emergency. But American support went far beyond simply permitting use of the base. The infrastructure of the base was improved and 12.5 million gallons of aviation fuel were made available, far exceeding the strategic reserves on the base. The Americans provided spare parts, Sidewinder air-to-air missiles, Stinger anti-aircraft missiles, Harpoon anti-ship missiles and Shrike air-to-ground radar seeking missiles.

Finally, American support in communication facilities and intelligence helped to fill a vital gap in Britain's capabilities. The US Department of Defence made available some of its military satellite channels, thus considerably improving communications between Fleet Headquarters at Northwood near London and the fleet after compatible equipment was purchased and installed. Britain also received SIGINT (signals intelligence) which had been passed on from listening posts belonging to the National Security Agency (NSA) to GCHQ in Cheltenham. Apparently the Americans had broken Argentine military codes – American sources claim that 98 per cent of intelligence about Argentine movements available to Britain was supplied by the US. Britain was also keen to obtain satellite pictures. This was technically difficult to obtain given that there was no coverage of the area by satellites with high definition reconnaissance equipment. At a later stage in the conflict the Department of Defence acceded to British requests and moved a satellite into an appropriate orbit to cover the Falklands area.[34] To what extent this contributed to Britain's military effort is still not clear.

Most analysts will accept that the strategic balance in the South Atlantic was close and that Britain could have conceivably lost the war. Unbeknownst to the world and even the British government, the United States would have stepped in the breach if Britain had faced military defeat. The concrete military assistance given did substantially aid the British military effort; the campaign would otherwise been have much more difficult, although not necessarily impossible to sustain with the same result. It had a limited effect on the political aspects of the conflict. The rumours to the effect that the United States were secretly aiding the British were sufficient to contribute to the souring of Argentine-American relations, but they lacked sufficient substance to enter into Argentine strategic calculations.

To conclude, the American decision to support Britain in the Falklands conflict of 1982 was not a foregone conclusion deriving from the 'special relationship' between the two. The decision was the outcome of a variety of factors, in the first place the calculation of US strategic interests. This issue was the most controversial in that there were voices in the administration which saw that US relations with Latin America were more important in these circumstances. The Europeanists prevailed in this instance not least because the survival of Mrs Thatcher in office was seen as critical to the success of American policy in the area of nuclear defence and arms control. Secondly, there were the perceived consequences of the Argentine intervention for the international system. The United States could not side with naked aggression in clear violation of international law, compounded by the refusal to abide by UN Security Council Resolution 502. Thirdly, the personal inclinations of some of the principal decision-makers were important too. Alexander Haig, a former SACEUR, saw himself as the defender and interpreter of European security interests in the Administration. For Ronald Reagan the personal relationship with Mrs Thatcher was of immense importance. The staunchly Anglophile Caspar Weinberger was prepared to back Britain up to the hilt both politically and by practical military aid.

What were the effects of Anglo-American co-operation on the course of the conflict? From a purely military point of view it is evident that American support by supplying intelligence, communication facilities and other equipment was crucial, if not decisive. From a political perspective, it looks different. Haig's mediation effort, albeit flawed, provided the opportunity for negotiation while the Task Force was sailing to the South Atlantic. The Americans did attempt to impress upon the Argentines that if a compromise was not found the British would go to war, the United States would have to back her, and she would surely prevail in any military confrontation. The junta was not convinced of any of these propositions. Buenos Aires might conceivably have been persuaded if the United States had supported Britain from the outset, together with overt military support. This might have brought home to the junta that its position was hopeless and that it would have to withdraw from the islands. The consequences of an Argentine withdrawal from the islands for a post-conflict settlement would have been very far-reaching, for if the United States had prevented a military conflict part of the settlement would have involved negotiations that would not have resulted in a return to the *status quo ante*, but would involve some of the ideas aired during the mediation effort. However, such a commitment entailed a very much greater political price than the Reagan administration felt it could pay.

Anglo-American co-operation during the Falklands Conflict strengthened the already close relationship between the Thatcher government and the Reagan administration. It is not evident, however, that the relationship would have developed very differently without the Falklands War. There was a basic agreement about fundamental political philosophy which was at the root of it. This did not prevent occasional strong disagreements on policy issues where British and American interests differed. The issue of sanctions against Poland and European involvement with the Siberian gas pipeline project was one such example prior to the Falklands Conflict; the following year, the US invasion of Grenada was strongly opposed by the British government.[35] The argument that the United Kingdom, as a matter of general policy, puts a cold analysis of its own interests, which include the preservation of unity with its partners in Europe, ahead of the Anglo-American relationship as a matter of general principle has been forcefully put by Jeane Kirkpatrick on the basis of an analysis of British voting patterns in the United Nations General Assembly and Security Council.[36] The only time when Mrs Thatcher felt obliged to support the United States against her better judgement, as a result of American support for Britain in the Falklands, was when she allowed the use of F-111 bombers based in the UK for the bombing of Libya. This decision was politically costly for her. In the context of the Reagan doctrine, military intervention was justified as legitimate means to advance a renewed effort to contain the expansion of Soviet power. The Thatcher administration, despite sharing much of the perspective of the US administration on the nature and objectives of Soviet foreign policy, and while recognising that the United States had a special role to play in global security, did not approve of this determined attempt to undermine the international consensus against the use of force. In the context of the debates about nuclear defence policy, the Thatcher government had to take into account that many Europeans found US foreign policy every bit as adventurous and dangerous as Soviet foreign policy. Furthermore the economic constraints on British defence capabilities forced a further shift towards the decline in out-of-area forces and towards emphasising the defence of Europe and the strategic nuclear deterrent, a trend weakened but not arrested by the Falklands War. For example, Britain was very reluctant to be drawn into in the Tanker War of 1987–8 but eventually became involved in arrangements for European national naval units to be sent to the Gulf through the Western European Union.

The collapse of Soviet power in Eastern Europe has led to a disengagement by the Soviet Union on a global scale. The awesome Soviet threat is no more. The argument for out-of-area capabilities to support the containment of the Soviet Union has become unsustainable. Whether similar capabilities are necessary for other purposes remains an open question. The Gulf crisis prompted by the Iraqi invasion of Kuwait in 1990 at least temporarily forged a new partnership between the United States and Britain on out-of-area responsibilities. How long that partnership endures will depend to a large extent on the long-term fall-out from that crisis and on future threats to international security in the post-Cold War world.

NOTES

1. Julius Goebel, *The Struggle for the Falkland Islands* (New Haven: Yale University Press, 1982) p. 442.
2. Paul Groussac, *Las Islas Malvinas* (Buenos Aires: Municipalidad de la Ciudad de Buenos Aires, 1936) p. 38; Goebel, p. 462.
3. Raymond W. Copson and Richard P. Cronin, 'The 'Reagan Doctrine' and its prospects', *Survival*, XXIX (1987) pp. 40–55. The author also acknowledges a private communication from Fred Halliday.
4. Alejandro Dabat and Luis Lorenzano, *Argentina: The Malvinas and the End of Military Rule* (London: Verso 1984) p. 49; see also Paul Eddy, Peter Gillman and Magnus Linklater, *The Falklands War* (London: Sphere Books 1982) pp. 59–63; Peter Calvert, 'Latin America and the United States During and After the Falklands Crisis', *Millenium*, 12 (1983) pp. 69–78.
5. Tam Dalyell, *One Man's Falklands* (London: Cecil Woolf, 1982) pp. 133 ff.
6. *Latin American Weekly Report*, 12 March 1982.
7. See Calvert, p. 72.
8. Quoted in Eddy, Linklater and Gillman, p. 124. Jeane Kirkpatrick later elaborated on this view, explaining that although she disapproved of the use of force, the case of the Falklands could be construed as a case of colonialism, that the 'international community' did not believe Argentina guilty of aggression and that Britain, in pursuing its case by force, could not argue that it was upholding international law or acting on behalf of the international community. See Jeane J. Kirkpatrick, 'My Falklands War and Theirs', *The National Interest*, 18 (1989/90) pp. 11–20.
9. It is widely believed that Haig's offer to act as mediator was an emulation of the example of Henry Kissinger's 'shuttle diplomacy' with the purpose of increasing his standing and power as Secretary of State and perhaps to earn himself 'a place in the history books'.

10. Max Hastings and Simon Jenkins, *The Battle for the Falklands* (London: Michael Joseph 1983) p. 104; Eddy, Linklater and Gillman, pp. 129 ff.
11. Alexander Haig, *Caveat: Realism, Reagan and Foreign Policy* (London: Weidenfeld and Nicolson 1984) p. 273.
12. Haig, ibid.
13. Haig, ibid. This remark was widely quoted in the press.
14. For further analysis, see Christoph Bluth, 'The British resort to Force in the Falklands/Malvinas Conflict 1982: International Law and Just War Theory', *Journal of Peace Research,* 24 (1987) pp. 5–20. This analysis of Thatcher's perspective, and in particular her preoccupation with the minimum achievement she would have to present to Parliament is confirmed to some extent by Haig, op.cit.
15. Haig writes:

    > In reporting to the President, I emphasized that unless some way could be found to alter British authority and provide for an Argentinean role in the government of the Falklands, Mrs. Thatcher's terms ruled out Argentinean acceptance. If Galtieri accepted her terms, it would be the end of him. *Just as Mrs. Thatcher must show that the junta got nothing for its use of force, Galtieri must be able to show that he got something.* (p. 274, emphasis added)

    Here Haig neatly sums up the central dilemma of the negotiations, but this insight is in stark contradiction to many of his other statements and his apparent behaviour during the negotiations.
16. Nicholas Henderson, 'America and the Falklands', *The Economist,* 12 November 1983.
17. Lawrence Freedman and Virginia Gamba-Stonehouse, *Signals of War* (London: Faber and Faber 1990) p. 181.
18. Freedman and Gamba-Stonehouse, ibid.; Haig, p. 283.
19. Haig, p. 284.
20. Haig, p. 285.
21. Haig, ibid.
22. For more detail, see Bluth, 'British Resort to Force', op.cit.
23. Haig, p. 290.
24. See Christoph Bluth, 'Just War Theory and the Falklands/Malvinas Conflict', *Journal of Peace Research,* 25 (1988) pp. 189–90.
25. This did not imply British agreement to a settlement on this basis. Nonetheless, Haig believed that if Argentina agreed to it the British would find it hard not to accept a settlement along such lines. At a later stage in the conflict Britain was more disposed to consider such a solution.
26. Freedman and Gamba-Stonehouse, p. 230.
27. Freedman and Gamba-Stonehouse, pp. 239 ff.
28. Calvert, p. 73; Freedman and Gamba-Stonehouse, p. 229.
29. For more detail, see Freedman and Gamba-Stonehouse, pp. 355 ff.
30. Desmond Rice and Arthur Gavshon, *The Sinking of the Belgrano* (London: Secker and Warburg 1984) p. 72.

31. Caspar Weinberger, *Fighting for Peace* (London: Michael Joseph 1990) p. 144.
32. *The Economist*, 3 March 1984.
33. In particular, Haig was kept in the dark. When stories about the extent of American military support surfaced in the press, Haig issued angry denials. Haig, op.cit.
34. This is confirmed by Weinberger, p. 152.
35. There was an interesting relationship between the Falklands conflict and the Grenada intervention. Britain's justification for military action in the Falklands had been based on the principle that, whatever the merits of the Argentine case with regard to their claim over the Falklands, the use of force was illegitimate and contrary to the most fundamental principles of international law. This was turned on its head by the United States in Grenada. The British Cabinet recognised without hesitation that an invasion would be against international law. After the event, Mrs Thatcher compared it to the Soviet invasion of Afghanistan.
36. Kirkpatrick, p. 16.

11 The Nature of Anglo-Argentine Diplomacy, 1980–1990

Guillermo Makin

It is a statement of the obvious to say that in the 1980s Anglo-Argentine relations underwent several changes. Despite them the danger of tension, even conflict, persists. The nature of the dispute has been only imperfectly addressed. In many quarters in both countries the return to diplomatic normality, coupled with the raising of trade restrictions and the lifting of the Military Exclusion Zone, has been welcomed as signifying a momentous change. But diplomacy must be the means to an end – the end being a compromise, perhaps embodied in the form of a treaty, that is politically acceptable to the elected governments of both countries.

To hold the binational political feasibility of any compromise might be taken as another statement of the obvious. Yet the politically obvious was overlooked many times in the 1980s by both Argentine and British politicians, especially during the three month run-up to the 1982 conflict. There is much to be said for caution following the ups and downs of the decade, but less to be said for easy optimsm after the re-establishment of relations between the two countries in 1989–90. However, few will argue with the rather commonplace proposition that the specifics of a settlement will take time to emerge. A compromise that is politically feasible in both polities, even after Mrs Thatcher's forced resignation in November 1990, can only be accelerated by a political decision that will only be taken if other internal political priorities beckon. A solution will require time, political leadership, and a multitude of binational exchanges and perhaps years of diplomatic activity. In the absence of a firm political lead, public opinion must be kept informed, whilst preserving the natural confidentiality of political and diplomatic exchanges, if the compromise is to emerge from the mist of conflict and prejudice. However, any settlement must satisfy one basic criterion: it must be politically acceptable to the relevant political actors in both political systems.[1] Islander opinion might be taken into account but though the islander lobby remains a factor, it is difficult to see it holding at bay a resolutely implemented British foreign policy.[2]

With the exception of unforeseen events, some of them seen as crises, Anglo-Argentine diplomacy in an issue such as the Falklands/Malvinas is largely conditioned by domestic political mores. Argentine and British foreign policies have returned to some of the elements of the pre-1982 period – trade remains at depressingly low levels[3] – but still many internal and external factors distract both political systems from the search for a durable, binational politically feasible compromise.

The Argentine political landscape of 1980–2 featured military unwillingness to purge criminal elements that had unleashed and implemented repression against the illegal left, as well as of the legal centre of the political spectrum.[4] At the time the military were able to resist political pressure, but towards the end of the decade they could only lobby like any other interest group, never certain that their views would prevail. The passing by Congress of a defence law disapproved of by the three services, the forced retirement of officers, the nature of Menem's staggered pardon, all indicate that the officer corps no longer holds the political prominence of the 1980–2 period. The process of re-establishing the dominance of elected authorities over the military is far from over, nor is it free from inconsistencies, but it is dramatic in creating an essential element of the atmosphere required for a compromise on the islands.

From 1983 to 1986 – Alfonsin's midterm – much could have been done to make an adventurous initiative on the islands but the opportunity was lost. Diplomatically, an open agenda with sovereignty not excluded but not actively discussed (the Alfonsin-Caputo formula) was scarcely distinguishable from Menem's sovereignty umbrella formula, but was still not capable of setting off the normalisation of diplomatic relations. The reasons for the difference are little understood but suggest much about the nature of Anglo-Argentine relations in the 1980s.

One of the features of the decade was an unprecedentedly dominant British Prime Minister reluctant to negotiate with the first elected Argentine President of the 1980s.[5] Indeed Mrs Thatcher was only with difficulty persuaded to send what turned out to be a haughty message congratulating Alfonsin on his inauguration. Her foreign affairs advisor after the conflict (Sir Anthony Parsons), when discussing with the PM possible diplomatic moves, was reportedly told that little could be done in the course of that parliament – that is until 1987 or 1988.[6] A few paraphrased Churchill in the hope that Thatcherite Britain would be magnanimous immediately after the conflict was over, but such hopes were not very realistic. Britain felt good in the new and unaccustomed aura of success after the Argentine surrender in June 1982. There could be little movement in what was left of 1982 and early 1983 because a general election was in the air in Britain, whereas in Argentina a democratic government had yet to be elected and seen to be firmly established; that is to have passed the very predictable ordeals of military turbulence and economic reconstruction.

One by one these very reasonable difficulties were removed. By the mid-1980s pride in British military success gave way to concern about the open-ended nature of the military commitment to the South Atlantic. However, given North Sea oil – and Argentines by 1985–6 became fully aware of this structural change in the British economy – Thatcherite Britain could afford 0.34 per cent of all state expenditure and 3 per cent of the defence budget in the 1982–7 period to be invested in the Falklands/Malvinas.[7] Perhaps the best explanation of the policy of the British government in the mid-1980s is that it seemed to be moved by a desire to send to Argentina messages that were substantially different to those sent in the 1960s and 1970s. As Baroness Young implied in her testimony to the House of Commons Foreign Affairs Committee, it was regarded as necessary to insist repeatedly that negotiations on sovereignty were not being contemplated by Mrs Thatcher's administration.[8] It appeared that the Argentine interpretations of the 1965–82 period could only be undone by throwing money at the problem and stationing ships, aircraft and men on a scale that worried most Latin America Ministries of Foreign Affairs. When the issue came up in Parliament the initial commitment to the 'paramountcy' of the wishes of the islanders came to be seen as unconstitutional. The Prime Minister became more careful except when sending her corny Christmas messages to the Falklanders, or when refurbishing her Iron Lady image after the politically damaging Westland affair in 1986; or in the same year on the occasion of the first visit of an Argentine parliamentary group to Britain since the conflict. In the end we may take it that paramountcy was discreetly dropped. However, the islanders retain a hold on British policy because they are few and seen to be 'the blood of our blood', as Mrs Thatcher put it in a BBC Panorama interview in April 1982. If and when this emotional hold is weakened, a compromise might become feasible. But then London will have to believe that it is more important to get on well with Argentina then it is to allow a population the size of a village to dictate British foreign policy.[9] That day did not arrive in the 1980s.

Alfonsin's inauguration in 1983 opened a first phase in the peculiar relations between Britain and Argentina featuring messages through the protecting powers.[10] Soon after being sworn in the Alfonsin administration seems to have proposed the re-establishment of diplomatic links, but, according to information leaked by Senator Adolfo Gass, the British response was regarded as unacceptable and unsatisfactory because it excluded any mention of sovereignty. Gass's revelation may be taken as a crafty way of informing public opinion in each country without involving the Argentine Foreign Ministry, or Alfonsin himself. Argentina reportedly stated its wish that contacts should lead to the opening of substantial negotiations, holding that in order to solve the dispute, sovereignty could not be ignored, acknowledging that due attention should be given to the interests of the inhabitants of the islands, as well as to the economic and scientific cooperation in the area, so as to return to the pre-conflict situation. Through the Brazilian ambassador, Britain proposed that trade and financial restrictions should be lifted, that arrangements for moving remains of Argentines from the islands or for visits of their families should be worked out, that air links between both countries should be re-established, that cultural, scientific and sports contacts should again be set up, and that the level of diplomatic relations should be raised. Argentina was said to have added new subjects regarded as indispensable: that the protection zone should be removed, that airport building should cease and that what came to be known as the 'Fortress Falklands' military concentration in the area should be reversed, that guarantees that nuclear weapons would be withdrawn from the area should be provided, and that the agenda should include the issues proposed by the UK.

It was reported that the Argentine government believed that the re-establishment of relations with the UK was extremely important. Alfonsin's proposal that a UN peace keeping force should guarantee the security of the islanders was repeated.[11] These leaks set out the positions of both governments very fully. They remained substantially the same throughout the Alfonsin administration, even though a number of the issues waxed and waned. Given the lack of importance of Latin America for the UK establishment, it is perhaps not surprising that British efforts to reach a compromise moved with the ponderous speed of a glacier.

A period of non-communication followed, punctuated by a series of more or less unofficial visits and return visits. That these visits should be necessary and newsworthy points to the lack of mutual knowledge in both political systems, perhaps the factor causing the 1982 conflict, as well as the lengthy stalemate that followed and the danger smouldering away even after relations improved in 1989–90.

The first visit was led by the Chairman and co-founder of the South Atlantic Council, Cyril Townsend MP (Conservative), plus George Foulkes MP (Labour) and Lord Kennet (Social Democrat). In Buenos Aires they met a much more high-powered set of parliamentary equivalents: Senators Adolfo Gass (Radical), Julio Amoedo (Popular Conservative) chairman and vice-chairman of the Senate Foreign Affairs Committee, respectively, and Senator Eduardo Menem (Justicialista) then not so well known or influential, as well as by the chairman of the Foreign Affairs Committee of the Chamber of Deputies, Federico Storani (Radical). The visitors reportedly stated that it was necessary to re-establish negotiations, tendering an invitation to visit London and mentioning the feasibility of long lease-back.[12] At the same time in June 1984 it was announced in Britain that the UK was prepared to re-establish relations whilst excluding sovereignty.

If the protecting powers assumed that the situation had changed enough to merit the organisation of more official Anglo-Argentine talks, it was a gross miscalculation. The Berne talks were preceded by speculation in the Argentine press arising from the visit to Buenos Aires of British parliamentarians, on confidence building measures, on the opportunities for an Argentine declaration of cessation of hostilities, all South Atlantic Council-sponsored moves that did not receive a very warm welcome in Buenos Aires, along with a lifting of the exclusion zone by the UK. The following options were also considered: UN trusteeship, joint sovereignty, autonomic status for Falklands/Malvinas within Argentina, and leaseback.

The newspaper *Clarín* revealed that talks were going on in Berne on 18 July 1984. Dante Caputo, the Minister of Foreign Affairs, confirmed that the meeting would take place, followed by a communiqué, simultaneously issued in Berne and London, mentioning a Swiss invitation to UK and Argentina to meet as from 18 July in Berne, the purpose being to facilitate an informal exchange of ideas without a predetermined agenda. Caputo warned that open talks could not exclude sovereignty. Once the talks were over came mutual recriminations that the Argentine delegation raised the need to discuss sovereignty and that the UK delegation supposedly responded that they were not prepared to discuss sovereignty. *Clarín's* diplomatic editor, Oscar R. Cardoso, wrote that Swiss sources assured Argentine diplomats that when sovereignty was raised, as it was agreed it would be given repeated statements by the new Argentine administration, the almost scripted UK response would be that they were 'not in a position' to discuss it. Argentina would then assume that sovereignty was not excluded, allowing discussion of other items. According to Cardoso and other sources involved in the talks, the UK delegation said that they were 'not prepared' to discuss it, which was taken as being quite different from the expected form of words. Sir Geoffrey Howe, then ensconced in the Foreign Office, echoed by all the British press, said that talks were interrupted because Argentina was not prepared to continue them on the allegedly agreed basis. Caputo informed the Chamber of Deputies that the prior agreement had been broken by the British delegation. The Swiss alleged that both parties were further apart than they, the Swiss, had imagined. The US Ambassador in Buenos Aires, Frank Ortiz, said that the US might intervene, taking the view that both countries were US allies, even though US intervention had not been requested.[13]

It is difficult to regard the talks as a failure inasmuch as the outcome conveyed the message that the Alfonsin administration would not negotiate for the sake of doing so and that negotiations, if they were to begin on a realistic footing, must not expressly exclude sovereignty. However, the end of the Berne talks closed the possibility of further direct government-to-government contacts until after the 1987 British general election. In fact, because of the unilateral British declaration of a provisional fishing zone, contacts were mediated by the US. Both the indirect and the indirect contacts, a feature of late 1983 and most of 1984, indicate that the two sides were far apart.

Visits by politicians not in power, as part of the learning process, were resumed when the Labour MP Donald Anderson, one of the party's foreign affairs spokesmen, travelled to Buenos Aires in September 1984. Argentine parliamentarians met Anderson who stated that the UK government, in his view, did not feel that ending the conflict was a priority, and that Mrs Thatcher wanted the re-establishment of links to proceed step by step, leaving other issues such as sovereignty for a later stage. Labour, said Anderson, took a different line, preferring the normalisation of relations to start from a realistic appraisal of all options that must include sovereignty.[14] Another meeting of parliamentarians followed, this time in Geneva. The Argentines warned that it was 'necessary' to include sovereignty. The British parliamentarians said that it was difficult for the UK government to accept the inclusion of sovereignty because that would entail ignoring the 1982 hostilities.[15]

The respective positions of each government were not very open to further change. Alfonsín stated his views to the UN: his administration's first priority was the debt, then came seven paragraphs on the 'Malvinas' (*sic*). The Argentine position, he held, had run into British intransigence. He repeated Argentina's willingness to negotiate, saying, yet again, that his government would only use pacific means, and that the re-establishment of relations with the UK was inconceivable if it was not certain that a mechanism would be set up that would allow a dicussion of sovereignty. It transpired that to Alfonsín a Hong Kong-type solution was acceptable but that the lease should run for a few months, not 99 years. Howe's response was picked up by the Argentine press – which was not always the case in the British press when the Argentines said something – that the UK was not prepared to discuss the sovereignty of the islands, and in an interview Howe predictably said that Hong Kong and other British territories were different.[16]

Signalling that in the future the Alfonsín administration would concentrate on the UN, Caputo restated the Argentine position, urging the UK to seek a negotiated solution that could assure a longstanding relationship between both countries and the rights of the islanders. Caputo asked rhetorically who benefited from the construction of the huge base on the islands ?[17] The UK representative replied that the future of the islands must be resolved by the islanders, and proposed that trade relations should be re-established, along with air services, and that cultural and sporting exchanges should be re-established as practical steps to improve relations but without negotiating sovereignty. When the votes were counted the Argentine motion totted up eighty-nine (including the USA), nine against and fifty-four abstentions, an increase of two on 1983. Abstentions included all of the EEC. Italy and France praised the Argentine proposal but justified their abstentions on grounds of European solidarity.[18] In the following years Argentine diplomacy eroded EEC support for Britain, both in terms of nations and significant allies supporting the UK motions. The results remained irrelevant inasmuch as the UK government would rather have had it another way but was prepared to live with UN disapproval. Argentina meanwhile had to travel down the UN road.

Having the US as a rather weary go-between was a further indication of what would be a feature of relations between both countries. The US Under-Secretary of State for Interamerican Affairs, L. Motley, said his government would try to see that the UK took a more flexible attitude.[19]

While Britain continued to insist on re-establishing normal relations, excluding sovereignty, the Argentine government took a quantum leap away from traditional Argentine foreign policy in its reliance almost exclusively on government-to-government relations. Foreign Minister Caputo was interviewed by Sir Nicholas Henderson for the BBC television programme 'Brass Tacks' and saw to it that an interview was published in *The Times* a few hours before the Falkland Islands Report was debated in the Commons. Bolder moves were yet to come: Alfonsin saw Neil Kinnock in Paris and later David Steel in Madrid, while in Madrid Alfonsin was interviewed by John Simpson for 'Newsnight' in October 1985. Thus the Argentine government established some contact with the British electorate and opened channels with the British opposition.[20] While more of the same was no doubt advisable, in the remaining three years of Alfonsín's term pressing domestic considerations made follow-up moves impossible.

A further irritant in this period was the news that First Land Oil and Gas was authorised by the government of the islands on 16 June 1984 to extract oil and gas. The predictable, perhaps the intended, result was a furore in Argentina, where many misguidedly believe that the islands are a repository of untold mineral wealth. The Argentine communiqué stated that no oil agreement would be recognised on Argentine territory in the 'Malvinas' and the surrounding area, and that measures such as the one taken by the islander administration severely imperilled normalisation of relations, urged by the UN. The Civil Commissioner, Sir Rex Hunt, argued that the islands never belonged to them, adding that he wanted diplomatic relations between Argentina and the UK but no links between the islands and Argentina. The UK sent a note to the UN Secretary General repeating that the islands were part of the UK and that the inhabitants had every right to start explorations for oil. For the UK measures such as these assured the economic development of the islands and were unconnected with normalisation of relations between UK and Argentina.[21]

Early in April 1985 Caputo said 'Fortress Falklands' would bring the East-West conflict to the southern hemisphere and criticised the UK for its refusal to discuss sovereignty. He took the line that fortifying the islands and placing one third of the UK fleet earmarked for NATO close to the islands, had no connection with defence because of Argentina's hypothetical threat. Caputo said yet again that the Mount Pleasant facility amounted to an aggression against Argentina and Latin America. Argentina called for a meeting of the OAS Council and Jorge Sábato, then the number two at the Ministry of Foreign Affairs, said investment, in what Argentines referred to as 'the base', had curious motivations because the Argentine government had fully discarded invading. Caputo in a speech in Washington to the OAS said 'the base' was a threat. UK preparations, he said, were out of all proportion, and British forces could even occupy strategic positions in continental Argentina. The OAS backed Argentina, with the US saying that it preferred a negotiated settlement. Caputo proposed that both countries should return to the situation preceding the conflict and negotiate the whole set of problems separating both countries.[22]

Analysis of the direct and mediated changes in 1983, 1984, 1985 and 1986 shows that there is a marked contrast with the swiftness of changes in 1989–90. Either a decision had been taken by the relevant elements of the British government that the first democratically elected government in Argentina could not at the early stages be trusted, or the gap between both polities was such that each side had to repeat positions a number of times so that the other took them partly or fully on board. Away from the shades of meaning perceived only by some specialists, the fact remains that the difference between an open agenda, coupled with some kind of agreement to deal eventually with sovereignty and a sovereignty umbrella is slight. Therefore the re-establishment of diplomatic links as a consequence of Menem's umbrella formula can only be explained by a political decision in London or by greater political familiarity after years of direct and vicarious bickering.

Perhaps the best indication that familiarity had by 1989 changed the political atmosphere, is the hostility in the British press, whipped up by Foreign Office off-the-record briefing, towards the visit in February 1986 by a high-level group of Argentine parliamentarians, when compared with the benign neglect towards visits in 1989 and 1990.[23] The 1986 visit confirmed that the consensus among British politicians on the most desirable policy towards Argentina and the islands no longer existed, reinforcing the possibility that movement in the form of more adventurous proposals could confidently be expected to begin sometime after elections in both countries in 1987 and 1989. Monitoring of all of the British media leads to the perception that portions of the British press were beginning to perceive and transmit that there is little truth in the caricature of the Argentine position spread by the Falkland Islands lobby and sometimes by the FCO: namely that Argentina merely wanted negotiations on the date of the full handover. In February 1986 both the *Guardian* and *The Times* picked up elements of a compromise framework.[24] In addition, the February 1986 visit confirmed a further constructive development in British politics: gone was the politically unwise willingness to talk only to English-speaking Argentines, or only to those with business connections with the UK who could only acquire power or influence because of a coup. Though the British FCO, very sensibly, agreed to meet almost any Argentine who saw fit to travel to London, British politicians seem to have realised that it was politically prudent to talk with those representing the consistent electoral preferences of the Argentine electorate.

With the unashamed use of hindsight it is clear that little could happen immediately after 1987 because in Argentina domestic politics made it difficult. The approach of presidential elections, added to economic and military turbulence, made it advisable to postpone the acceptance of any Argentine proposals till a new Argentine president was in power. The contrast with the mutual ignorance of 1982 could not be more marked. Something had been learned about Argentine politics. In 1982 Mrs Thatcher, with little knowledge of Argentine politics, attempted to deter Galtieri from sending forces to the islands by asking President Reagan to intervene, a sure way of obtaining the opposite result. Ignorance was not one-sided; the Argentine military and their advisers knew nothing in 1982 of Thatcher's political personality. Arguably the change was subtle and remains to a large extent unreported. Before 1982 British politicians knew little of the South Atlantic dispute, less of Argentine politics. Argentine ignorance of British politics was even more abysmal. When and if they put their minds to it, British politicians and occasional observers deluded themselves with visions of Argentine parties or officials compliant with perceptions of British interests, or, in a further flight of fancy, keen to surrender the Argentine claim. After the conflict the two reports by the House of Commons Foreign Affairs Committee, plus improved coverage by the British media of Argentine views, arguably caused the few British politicians interested in the issue to realise that the Argentine claim would not go away.[25] Some have even gone so far as to say that sovereignty will of course have to be included later, this being said privately, though more daring politicians, such as George Foulkes, have said so in public.[26]

This mostly private acceptance that eventually sovereignty will have to be an item in any negotiations, though minor and yet to gain much influence, let alone power, has its promise for Argentine ears. Again as in 1966–82 talks on sovereignty appear as eventually possible in the long term, despite repeated statements by the British government to the contrary. Argentines construe the tension between private views and public statements as meaning that the dispute does not seem at the dead end the military wanted it to be in 1982. As was the case immediately after the 1982 conflict and before a democratically elected administration was installed, Argentine politicians have other priorities in the 1990s. Argentine foreign policy until 1989, when hyperinflation caused Alfonsin to resign early and Menem was rushed into power, has suffered from excessive reliance on the UN. However, the size of the majority supporting progressively milder Argentine-sponsored resolutions showed the limits of succeeding in isolating Britain. By October 1989 the Menem administration had realised the limitations of the UN.[27]

The genesis of the Menem umbrella formula lies in a series of external and internal factors. The unilateral declaration of the exclusive fishing zone by the UK and the 'Fire Focus' military manoeuvres were the main external developments that drove home the message that the UK under Mrs Thatcher was in earnest about staying and transforming the islands. The Argentine establishment realised that little could be done other than raising the issue with the UN and the OAS. Internally, continued economic malaise further convinced Argentines that their country would remain far from able to do anything. By October 1988 Menem was campaigning for the presidency and many close to him believed that 'something had to be done about the British'. Many thought that Menem would be warlike, given the garbled reporting of a speech in Tierra del Fuego, but they were wrong. Peronist thinking is characterised, among other things, by pragmatism. On this issue, as in many others, Menem moved ahead of those Peronists who looked back at the 1940s and early 1950s through the pink lens of nostalgia. When the 'something to be done' turned out to be the umbrella formula, this time the UK responded with astounding speed, perhaps because as talks unfolded in 1989 and 1990, the Foreign Office became increasingly confident that it had secured most of its objectives.[28]

If developments in the 1980s have taught each political system something about the nature of the other, progress towards a compromise may be made before the century is over, only then placing relations between Argentina, the UK and the islands on a sure footing. How much has been learned? Many are determined that there will be no repetition of the conflict of 1982. A few believe it is both desirable and inevitable that there should be moves towards a compromise that is politically feasible in both countries. This indulgence in optimism over the feasibility of a settlement also results from the political changes in Argentina and the toppling of Thatcher in Britain. Domestic political developments, of which the outcome of elections are the main feature, may provide politicians and officials with confidence enough to move, if in addition forceful leadership is deployed, to consolidate the acceptance by public opinion of a lasting, mutually acceptable compromise. Windows of opportunity may present themselves, but will Britain bother and will Argentina be ready?

NOTES

1. Giovanni Sartori, *Parties and Party Systems, a Framework for Analysis* (Cambridge: Cambridge, 1976) pp. 121–2. Relevant political actors are those who can make and unmake governments. This ability is determined by electoral strength, coalition potential or by ability to blackmail.
2. Even before the 1983 presidential elections in Argentina leaders of the democratic political spectrum recognised that the islanders would require, amongst other things, cast-iron guarantees. Cf. G. A. Makin, 'Argentine Elections and Items for a Negotiation with the UK', *Bulletin of Latin American Research,* 3 (1984).
3. Argentine direct exports to the UK were £66.8 million in 1988. British exports to all of Latin America added up to only 1.70 per cent of all British exports in 1987 and to an even more depressing 1.57 per cent in 1988.
4. For a full development of the evidence based on government releases, *La Prensa, La Nación* and *La Opinión,* G.A. Makin, 'Political Crises in Argentina: 1955 and 1975–76', University of Cambridge PhD dissertation, 1985. For violence aimed at the centre of the political spectrum, idem, 'Argentina, the Authoritarian Impasse', in C. Clapham and G. Philip (eds), *The Political Dilemmas of Military Regimes,* (London: Croom Helm, 1985), especially p. 169.
5. Lawrence Freedman, *Britain and the Falklands War* (Oxford: Basil Blackwell, 1988), p. 67.
6. Interview with John Simpson, reported in G.A. Makin, 'Los Partidos Politicos Británicos y sus Posturas sobre Malvinas', *América Latina/ Internacional,* 3 (1986) and, 'Malvinas: Las Negociaciones Secretas', *La Semana,* 28 January 1987.
7. *Clarín,* 24 March 1985 and 27 April 1987.
8. *Second Special Report from Foreign Affairs Committee, Session 1982–83, HC 378, Minutes of Proceedings, Session 1982–83,* HC 380, Minutes of Evidence, (London: HMSO, 1983) which recommends leaseback in the body of the report in material that was to be part of the conclusion but was aborted because of the 1983 election. *Fifth Report from the Foreign*

Affairs committee. Session 1983–84, Falkland Islands vol I *Report and Minutes of Proceedings* and *List of Appendices to the Minutes of Evidence*, vol II (London: HMSO, 1984) is a much more timid document, stating that eventually an arrangement must be reached and hinting at the need to wait and see if Argentine democracy stabilised.

9. Cf. interview with Jim Lester MP, in *Clarín*, 26 April 1984.
10. *La Nación, La Prensa, Clarín* 11 December 1983 Thatcher-Alfonsin, messages: '*donde hay voluntad hay solución*' (where there is a will there is a way) replied Alfonsin, prophetically saying that during his period in office the Thatcher government would not have the will.
11. *Clarín* and *La Nación* 9, 14 and 15 April 1984.
12. *Clarín* , *La Nación, La Razón, La Prensa* 26 and 30 June 1984.
13. *Clarín*, 3 July 1984; *La Prensa*, 5 July 1984. *Clarín* reveals talks in Berne, 18 July 1984, *La Razón* 18 July 84 and *Clarín, La Nación, La Prensa, La Razón*, 19 July 1984. *La Razón*, 19 July 1984 and *Clarín, La Nación, La Prensa* 20 July 1984. *Clarín, La Nación, La Prensa*, 20 July 1984 and 21 July 1984 on what the Swiss had to say, also *Ambito Financiero*, 20 July 1984. On statement by US Ambassador, Frank Ortiz, *La Prensa*, 21 July 1984.
14. Cf. interview with D. Anderson, *Clarín* 9 September 1984; *La Nación, Clarín* and *La Prensa*, 16 September 1984.
15. *Clarín, La Nación, La Prensa*, 27 and 28 September 1984.
16. *Clarín, Cronista Comercial, La Razón*, 25 September 1984, on Argentine position. For Howe's response cf. *Clarín*, 27 September 1984.
17. *La Nación, La Prensa,Clarín*, 1 November 1984
18. *Clarín*, 2 November 1984.
19. *Clarín, La Nación, La Prensa*, 21 November 1984.
20. 'Brass Tacks', BBC TV, 12 December 1984; *Clarín*, 5 December 1984.
21. *Noticias Argentinas*, 2 January 1985; *Clarín, La Prensa, La Nación*, 3 January 1985.
22. *Clarín, La Prensa, La Nación*, 7 April 1985; 5, 14 and 17 May 1985.
23. *Clarín*, 14, 15, 16, 17 and 24 February 1986; *The Times, The Guardian, The Daily Telegraph* and *Financial Times* for the same dates.
24. *The Guardian*, 20 February 1986, a magisterial piece by Arthur Gavshon, perhaps the only British journalist to have held a lengthy conversation with the Argentine group and to have reported it without adulteration. The rest of the British press, as usual, responded to governmental briefings. *The Times*, 22 February 1986, quoted quite representative views from Senator Amoedo.
25. *Foreign Affairs Committee, 1982–83* and *1983–84 Reports.*
26. *The Times*, 20 May 1986. 'Labour move for Talks on Falklands' is the headline. 'Preparations to enable a Labour government to reopen negotiations with Argentina about the future of the Falklands, including sovereignty, have been recommended by an internal party report'. For the inevitability of sovereignty as an item for future negotiations, interview with current member of House of Commons Foreign Affairs Committee, September 1989.
27. Interview, Eduardo Menem, 5 September 1989.
28. Interview, Carlos S. Menem and members of his team, 4 November 1988.

12 Public Attitudes and the Future of the Islands[1]

Felipe Noguera and Peter Willetts

A small group of islands in the South West Atlantic is known in the Spanish-speaking world as the Malvinas and in the English-speaking world as the Falkland Islands. They are nearly 1,000 nautical miles from Buenos Aires, but less than 200 miles from the southern Argentine province of Tierra del Fuego. The temperate climate sustains an economy that used to be based on sheep farming. Since 1986 the main income has been derived from licensing international fishing fleets, in a Falkland Islands Interim Conservation and Management Zone (FICZ) covering the waters around the islands. There are just 2,000 permanent inhabitants of British descent, with another 1,000 expatriate residents, consisting mainly of the British forces at Mount Pleasant, plus small numbers of business people involved in the fishing industry and officials employed in government services. In another chapter in this book Peter Beck examines the nature of the claims to the islands made by the Argentine and the British governments. The British victory in June 1982 concluded the conflict for the immediate future, but the underlying dispute still remains.

In the 1960s it appeared that Britain might transfer sovereignty over the islands to Argentina. The combination of reactions from the islanders, from Parliament and from the media led the British government to reject the draft agreement. A similar process occurred in 1980 when negotiations between Britain and Argentina moved towards the idea of a 'leaseback' agreement. Under this proposal the British would have recognised Argentina's sovereignty claim, but the Argentinians would have agreed to allow the British to administer the Islands for a fixed period of years. After being severely attacked in Parliament, the scheme had to be withdrawn. Since the ceasefire in 1982, a debate on the future of the Falklands/Malvinas Islands has been under way in many forums. Clashes between the Argentine and the British governments' positions have been evident at the annual debates of the United Nations General Assembly and in the two mini-crises over the declaration of the fishing zone around the islands in October 1986 and over the British 'Fire Focus' military exercises in March 1988. At the same time there has been an implicit desire by both sides for relations to improve, as evidenced by the attempt at talks in Bern in July 1984; the British lifting of trade restrictions in July 1985; President Alfonsin's conciliatory proposals of November 1986; the exchanges via Washington in 1987–8; President Menem's offer to put sovereignty 'under an umbrella' in July 1989; and the talks in Madrid that led to the re-establishment of diplomatic relations in February 1990.

The debate about the future of the islands has produced a variety of creative solutions.[2] In both Britain and Argentina lease-back still has some adherents, notably among active and retired diplomats.[3] In Argentina the emphasis has usually been upon recognition of the Malvinas as an Argentine province, with it being granted a special autonomous constitutional status, so that the islanders could maintain control over their way of life.[4] In Britain the idea of an autonomous province, comparable to the Aland Islands with a Swedish population having autonomy under Finnish sovereignty, has been explored by the South Atlantic Council.[5] Some of the more internationally-minded have advocated the territory coming under the UN Trusteeship System.[6] Martin Dent, from the University of Keele, has been a persuasive advocate of Argentina and Britain sharing sovereignty over the islands and leaving the day-to-day decision-making to a Governor and a local Council.[7] Edna Lemle, a New York business woman with a commitment to conflict resolution, has proposed a 'pentocracy' for the islands, consisting of five elements; the two external disputants, the local community, business interests and the UN as the voice of the international community. This scheme is designed as a model for resolving many regional disputes.[8] Occasionally there are hints from the islands that independence might be a desirable option, though that would be meaningless unless it were either accepted by the Argentine government or guaranteed by a continued British military presence.[9]

At the official level and among private groups, both in Argentina and in Britain, the debate about the future of the islands has been within a small elite, consisting of individuals who each have very personal reasons for the issue being salient to them. This chapter examines opinion poll data in Britain and Argentina to see how the general public in each country responds to the concerns of the two elites. First there is an analysis of public attitudes to the improvement of bilateral relations between Britain and Argentina. Secondly, it is important to know whether attitudes towards changes in the *status quo* are inflexible or whether a negotiated settlement to the Falklands/Malvinas dispute might be acceptable. Then the chapter changes direction by considering the issue from the perspective of the two publics, to see whether they differ from the elites.

In their normal coverage of public attitudes in Britain to a wide range of issues, the Gallup Poll asked a limited number of questions on British-Argentine relations on three occasions in July 1984, October 1984 and February 1986. In addition a wider range of questions was asked in March 1990. Meanwhile in Argentina, SOCMERC (Mora y Araujo, Noguera and Associates) have held a regular series of polls and in March 1990 they were commissioned by the South Atlantic Council to ask the same questions of the Argentine electorate as Gallup was asking that month in Britain. Thus the March 1990 polls in each country give a basis for direct comparisons of attitudes in the two countries. A further poll in Argentina in August 1990 updated some of the March 1990 results.[10] (See Table 12.1 for details.)

If we consider first of all what were British attitudes to five specific steps that might be taken to normalise relations, we find there was overwhelming approval for better relations between Britain and Argentina and no suggestion that the breaks

TABLE 12.1 *Percentage approval of steps to improve relations with Argentina (figures in brackets give the percentage who said 'strongly approve')*

	July 1984	Oct 1984	Feb 1986	March 1990
Resume trade	69 (19)	65 (10)	–	68 (14)
Allow direct air flights	–	–	–	69 (13)
Agreement on fishing	69 (18)	70 (10)	62 (11)	74 (19)
Diplomatic relations	76 (22)	74 (12)	70 (13)	70 (17)
Visits to war graves	94 (42)	92 (31)	87 (26)	90 (36)

caused by the war should have continued. Those interviewed were asked whether they would have approved or disapproved 'if the government decided to do the following: resume trade between Britain and Argentina; allow relatives of dead Argentinian soldiers to visit the graves in the Falklands; reach an agreement with the Argentinians on fishing in the seas around the Falklands'; or 'resume diplomatic relations between Britain and Argentina'.[11]

It is striking that on each occasion more than 60 per cent of the British electorate said they 'approve' or they 'strongly approve' of each measure, while less than a quarter said they 'disapprove' or 'strongly disapprove'. Although one might have thought it was the least sensitive step to take, the resumption of trade relations received the least approval and the most disapproval. Support for a fishing agreement fluctuated, with a dip in February 1986 and an increase in March 1990, for which there seem to be no obvious explanations. The establishment of diplomatic relations received somewhat higher levels of approval, though there was less approval when relations were re-established after eight years than in the earlier period shortly after the war. The variation in responses to these three steps are essentially little more than sampling fluctuations against a uniform pattern of high approval. When we come to the question of whether the public would approve if the government decided to allow visits to the war graves, the difference was dramatic: approval for the visits was absolutely overwhelming.

The results do not show any significant variation between Conservative and Labour voters. This is contrary to what might have been expected from the idea of a 'Falklands factor' sustaining support for the Conservative Party. A long debate has raged in the academic journals between those who argue that the Conservative government's popularity increase in 1982 can be explained by the economy moving out of the 1981–2 recession, and those who argue that there was in addition a substantial benefit to the Conservatives from public support for fighting and winning the Falklands War. What is common to all those in the debate is the view that unemployment made the government unpopular in 1980–1 and continued to be important during and after the 1982 war. It also must be agreed that the Falklands factor emerged in May and June 1982 and declined thereafter. (Some of the early writers, using data for 1982–3 only, gave the impression that the Falklands factor was 'permanent', but that should now be interpreted as 'a substantial effect continuing until the 1983 general election'. The Conservative government recorded in early 1990 the lowest level of support since polls began, so there was not a truly permanent effect.) A team from Essex University have argued that the government was only boosted by 4 per cent in the polls and this effect evaporated rapidly. Another scholar has estimated that the gain was 11 per cent in June 1982, decaying to 5 or 6 per cent in the June 1983 election, and becoming less than 1 per cent by the end of 1985. Either way the Falklands factor was no longer significant in the late 1980s and thereafter.[12]

Nevertheless, to say that the Falklands factor ceased to explain support for the Conservative Party, when the war was supplanted by other issues, does not automatically mean that there will be no party variation in attitudes on questions that invoke the Falklands dispute. Yet, in practice, memories of the war or loyalties to Mrs Thatcher scarcely differentiated the attitudes of Conservative supporters from those of Labour supporters towards improving relations with Argentina. There is a different sense in which a Falklands factor was undoubtedly of importance in British politics: Mrs Thatcher herself gained great self-confidence from the success of the Task Force, and continued to maintain a strong personal interest in all aspects of policy on the Falklands as long as she remained in office. With her resignation as Prime Minister, this aspect of the Falklands factor also disappeared, and the Foreign Office ceased giving such high priority to the Falklands question.

Nor is there any variation in attitudes when the results are broken down by the sex or the age of the respondent. The first impression from the results might be that men were more in favour of improving relations than women were. However, women were markedly more inclined to answer 'don't know'. When the percentages are recalculated excluding those who said 'don't know', the differences become minimal. There was some impact of class in the March 1990 survey, though less in the previous ones. 'Strong agreement' with the measures was about 10 per cent less in the lower D/E class category than in the higher A/B/C1 category. In the July 1984 survey, Gallup also asked the voters a battery of eighteen questions about Mrs Thatcher's image. Again, if Mrs Thatcher's personal standing

TABLE 12.2 *Party choice and support for improving relations, March 1990*

Percent that 'Agree strongly' or 'Agree' with each point	*UK Total*	*Cons*	*Lab*	*Argentine Total*	*PJ*	*DP+ Rad*	*UCD*	*Left*
Ending trade restrictions	68	66	70	75	76	79	93	63
Direct air flights	69	67	69	78	73	85	93	74
Fishing agreement	74	73	75	71	74	73	71	63
Diplomatic relations	70	69	70	81	81	79	89	67
Visits to war graves	90	88	91	97	95	97	96	94

NOTE: Figures are given for the main parties: Conservative; Labour; *Partido Justicialista* (Peronist); *Unión Civica Radical* (Radical); the Democrat Party and the Union of the Democratic Centre (right wing); and the Movement for Socialism, the *Intransigentes* and other left parties. In both countries there were also minor parties and those who made no choice.

had been enhanced permanently by her leadership during the Falklands war, attitudes to questions on Argentine-British relations might be expected to have related to whether she was accorded a positive or a negative image. None of the eighteen questions on Mrs Thatcher's image correlated even minimally with any of the four questions on Argentine-British relations.[13]

Attitudes in Argentina in March 1990 were very similar to those in Britain: all the measures to improve relations received majority support, with trade, direct air flights, diplomatic relations and visits to the war graves each obtaining 7–11 per cent more support in Argentina than in Britain. Only on the question of reaching 'an international agreement, including Britain, Argentina, Uruguay and the other fishing countries, to manage and conserve the fish in the South West Atlantic' did the level of Argentine support fall below the British level, and then only marginally. The difference on this point may be explained by the lower saliency of the economic and environmental aspects of the conservation of fisheries in Argentina. As in Britain, there were no significant differences between the two main parties, the *Justicialistas* (commonly known as the Peronists) and the Radicals. (There was just one curious anomaly: that a greater proportion of the Radicals wished to see direct air links.) Nor were there differences in attitudes based on sex or age, except that again women were slightly more likely to reply 'don't know'. In contrast to Britain, the results for Argentina showed a definite effect of class upon responses. There was a very big increase in the 'don't know' reply in the lower class category, reaching 27 per cent on trade, 35 per cent on fishing and 38 per cent on air flights. This inevitably reduced the proportion in the approval categories. Nevertheless, at least half of the lower class approved of each of these measures.

When we consider the minor Argentine parties, there are distinct differences in the responses. The right-wing parties, the Union of the Democratic Centre and the Democrat Party, which support free enterprise and a greater integration with the West, but also number some nationalists among their supporters, managed to produce even bigger proportions approving three of the measures. Their supporters showed 8 per cent, 15 per cent and 18 per cent more than the national average, approving of resuming, respectively, diplomatic relations, air flights and full trade relations. On the other hand, supporters of the Movement for Socialism, the *Intransigentes* and other small left-wing parties have been both strongly anti-business and highly nationalist. Their supporters showed a lower than average proportion approving of any of the measures, particularly being 8 percent down on a fishing agreement, 12 per cent down on trade and 14 per cent down on diplomatic relations. Nevertheless, considered on their own, the figures for the hard left in Argentina still show high approval for all the measures to improve relations between the two countries.

Thus, neither the British government nor the Argentine government has anything to fear from any sector of their electorates. All the variation in attitudes towards improvements in bilateral relations is simply between high levels and overwhelming levels of approval. The results are particularly striking when one considers the points that might have evoked a more emotional response. In both countries humanitarian concerns for the families that suffered bereavements during the war override nationalist antagonisms that can be invoked over soldiers' graves. In Britain, media coverage of the visit to the Falklands by two members of the Gimenez family in October 1986 for the funeral of Lt Miguel Gimenez displayed some of the tabloid press at their worst.[14] A few islanders then and since have expressed misgivings about Argentine people coming to the Islands, even for the purpose of visiting the war cemetery.[15] Similarly in Argentina the Gimenez family received only discouragement from the government over their desire to go to the funeral, and the organisations for the relatives of the war-dead have been caught up in political divisions. Soon after the 1982 war a nationalist figure, Sñr Di Stefano, tried to exploit the question by taking a ship full of relatives to challenge the British exclusion of Argentine shipping from Falklands waters. Public opinion as measured in these polls shows that such excesses need have no bearing on policy.[16]

It is worth noting that the evidence from the opinion polls that there need be no worry about public opinion affecting direct Argentine-British bilateral relations is also substantiated by the range of bilateral contacts that began without any problems in the mid-1980s. Religious and academic contacts developed smoothly in the early years and cultural and sporting exchanges soon followed (with the exception of polo). Trade and other commercial activities took longer, both because of economic difficulties and because of official impediments, but much effort has been put in by business people as soon as diplomatic exchanges led in 1989 to official encouragement of trade. As early as June 1984 a South Atlantic council delegation of three MPs successfully made a wide range of contacts in Buenos Aires. When the return visit of the Argentine Congressional delegation came to London in February 1986, fears that there would be hostile press reactions proved to be groundless, and the delegation was very well received. Since the initial visits, many politicians have made short trips in each direction without the events being considered at all newsworthy. The first direct governmental contact occurred at the UN in New York between Sñr Dante Caputo, the Argentine Foreign Minister, and Sir Crispin Tickell, the British Permanent Representative to the UN, in December 1988, cautiously, but with no controversy, and now governmental contacts are part of the normal flow of diplomacy. At no point has there been significant public reaction in Britain or Argentina against these developments.

If there are no serious problems impeding the improvement of direct Argentine-British bilateral relations, what is the situation with respect to the Falklands/Malvinas sovereignty dispute? The first question about the dispute asked in the surveys covered the possibility of leaseback being acceptable as a settlement (see Table 12.3). Respondents were told 'Before the war Britain and Argentina discussed the idea of "leaseback", that is Britain

TABLE 12.3 *Response to the idea of leaseback for the Falklands*

	UK July 1984	UK Oct 1984	UK Feb 1986	March 1990 simultaneous surveys UK Total	Cons	Lab	Argentine Total	PJ	Rad	DP+ UCD	Left
Very good idea	8	9	8	6	4	8	4	2	5	4	2
Good idea	30	29	26	23	19	28	28	29	25	54	33
Bad idea	33	27	32	36	35	35	47	41	49	36	48
Very bad idea	15	15	13	17	28	12	–	–	–	–	–
Don't know	14	19	21	19	14	17	21	28	21	7	17

would give the Falklands to Argentina but the islanders would remain under British control for the lifetime of the present islanders'. They were then asked 'Do you think that this would have been a very good, good, bad or very bad idea?'. In all four polls in Britain there was a majority against leaseback, with the majority becoming slightly higher in the March 1990 survey. While there is very little difference by sex, age or class, we do now find an attitude difference between the parties. Nearly two-thirds of the Conservatives in March 1990 rejected leaseback, compared to only half of the Labour supporters (a 16 per cent difference in the proportions saying it was a 'bad' or 'very bad' idea); in the previous survey the party difference had been somewhat more pronounced (a 22 per cent difference in February 1986). The results in Argentina in March 1990 were relatively similar to those in Britain, except that the right-wing did produce a majority favouring leaseback. The only marked difference in Argentina, by sex, age and class, was in the tendency for more women, older people and the lower class to say 'don't know'.

In the 1984 and 1986 surveys in Britain, other possibilities for a settlement of the sovereignty dispute were explored by asking 'What do you think would be the best long-term future for the Falklands – to remain permanently British, to be handed over to the Argentinians with suitable guarantees, to be under lease-back, to be under joint control of the two countries, or to be under a United Nations administration?'. The results (see Table 12.4) show there was high continuity between the three polls, with remaining permanently British being the preferred option, and UN administration as the compromise receiving

TABLE 12.4 *Options for the future of the Falklands*

	July 1984	*Oct 1984*	*February 1986 sample Total*	*Cons*	*Lab*	*Lib/SDP*
Permanently British	37	37	38	51	32	34
Handed to Argentina	9	12	10	4	17	10
Lease-back	5	6	6	6	4	8
Joint control	15	13	13	9	16	11
UN administration	24	21	19	18	16	22
Don't know	10	11	14	11	14	14

the most support. This could be interpreted as the British public desiring to maintain 'Fortress Falklands' and rejecting change in the *status quo*, until one notes that less than 40 per cent chose the 'permanently British' option. The four options involving a change in the current situation and some possibility of a settlement with Argentina together were endorsed by about half the electorate (just over half in 1984 and just under half in 1986). It is not surprising that at this point we find the strongest differentiation in attitudes between Conservative supporters and those from the other parties. Nevertheless, the Conservative supporters are *not* united in wishing to maintain sovereignty: 44 per cent of the Conservatives in October 1984, 37 per cent in February 1986 and 44 per cent again in March 1990 were prepared to endorse giving up British rights to the islands, by choosing one of the compromise options.

These surprising findings were explored in more detail in the March 1990 survey, by asking separate questions about the possible compromise outcome (see Table 12.5). The first question simply asked for a choice between four outcomes: the two official government positions, evading the conflict by giving the Falklands independence or a general unspecified compromise arrangement. The results in Britain were close to the three previous surveys, except that the new option of an an independent Falklands gained substantial support. Only a third of the British electorate supported the official British government position of keeping the islands permanently British. The same question produced results in Argentina that were rather more in the pattern one might have expected. Even so, a total of 28 per cent of the Argentine population were prepared to choose

TABLE 12.5 *British and Argentine willingness to compromise, March 1990*

What do you think would be the best long-term future for the Falklands – to remain permanently British, to be handed over to the Argentines, to be independent, or some compromise arrangement?

	UK Total	*Cons*	*Lab*	*Argentine Total*	*PJ*	*DP+ Rad*	*UCD*	*Left*
Permanently British	33	48	29	2	1	1	7	4
Handed to Argentina	9	5	13	66	75	64	54	63
Be independent	22	20	19	8	5	8	11	11
Some compromise	25	19	27	18	11	24	25	17
Don't know	11	8	11	6	7	4	4	4

an option other than the Malvinas coming under the Argentine flag. The nationalist traditions of the *Justicialistas* and the pro-British traditions of the right-wing show clearly in the party differences on this question. In both countries the largest group supported the official position, but the British electorate were much less solidly behind the maintenance of British sovereignty than the Argentine electorate were behind their government's claim. Interestingly, in both countries one-tenth of the people had no commitment at all to the sovereignty dispute: 9 per cent of the British were willing to hand the Falklands to Argentina, while 10 per cent of the Argentines were willing to leave the islands with Britain or to accept that they could be independent. Again sex, age and class make little difference in either country, with just one exception, that acceptance in Argentina of independence for the islands drops from 16 per cent in the youngest group to 3 per cent and 6 per cent in the two oldest groups.

The possibility of a compromise settlement looked much stronger when various possible options were put directly to the voters, but people in the two countries had very different ideas on which compromise they would make their first choice (see Table 12.6). The Argentines overwhelmingly preferred simply to give guarantees to the islanders, so that their way of life could be preserved. But with this reassurance the proportion of British voters willing to hand over the Falklands only went up from 9 per cent to 13 per cent, the increase being the same for both Conservative and Labour supporters. Neither side gave more

TABLE 12.6 *British and Argentine preferred compromise, March 1990*

Various compromise arrangements have been suggested. Which of the following do you find the most acceptable?

	UK Total	*Cons*	*Lab*	*Argentine Total*	*PJ*	*Rad*	*DP+ UCD*	*Left*
An Argentinian province with guarantees of a separate way of life for the Islanders	13	8	16	75	79	75	68	78
Leaseback	6	4	7	1	0	1	0	0
Joint control of the two countries	13	8	16	11	9	13	14	13
A United Nations administration	42	48	42	3	2	1	14	4
None, Don't know	26	32	20	10	9	11	4	4

than minimal support to leaseback, while the 'perfectly-balanced' compromise of joint control gained the support of an eighth of the voters in both countries. The totals were made up by the British giving the highest support to the idea of a United Nations administration, but this obtained minimal endorsement from Argentines.

Unfortunately, the question did not translate to have an equivalent weight in the two political cultures, because giving the Falklanders guaranteed rights within an Argentine province is much easier for the Argentines to accept than any of the other compromises. It would have been better had this option at least referred to autonomous status within Argentina, guaranteed by a United States or United Nations military presence. Nevertheless, even when offered a 'perfect compromise', fewer Argentines made their first choice the Falklands being a province of Argentina than earlier in the same survey had supported direct air flights to Britain, re-establishment of diplomatic relations or visits to the war graves. The figures did not leave much room for party differences to emerge. The March 1990 figures appear to show the right-wing giving higher support than the other groups to a UN administration, but this finding was not repeated in the August 1990 survey. The party differences in Britain did appear, but not in a major way:

Labour supporters were somewhat more likely to opt for joint control and Conservative supporters were somewhat more likely to choose a UN administration.

The next four questions constrained the respondents even further, by forcing them to say separately for each compromise option whether they would find it acceptable. Although the questions thus bias the respondents towards serious consideration of some of the possible compromises, they were free to reply that they did not find them acceptable. This is a realistic way to put the questions before the electorates. If the Argentine and British governments were to negotiate a settlement at some time in the future, the electorate would simply be faced with accepting or rejecting the agreed compromise, by voting in a general election or perhaps in a referendum in Argentina. On these questions the reactions to leaseback and to joint control again showed no significant differences between the voters in the two countries. Each of the two potential compromises was rejected by clear majorities in both countries. The party differences did appear, as before, but only in the case of the Argentine right-wing has the majority turned round to provide acceptance of joint control.

The British voters continued to show that a UN administration was their preferred compromise, particularly among Labour supporters. This time, when considered as a separate question, acceptance of a UN solution in Britain gained a two-to-one majority, both in the electorate as a whole and among Conservative supporters, almost reaching a three-to-one majority among Labour supporters. After Mrs Thatcher, any other British Prime Minister has a good chance of being able to sell a UN administration to the electorate as an acceptable compromise settlement of the sovereignty dispute. Unfortunately, it is not possible to make a proper comparison between the British and the Argentine publics' attitudes to the UN, because due to an administrative error this question was not put in Argentina in the March 1990 survey. The results in Table 12.7, for just the UN question, are from the August 1990 survey. Nevertheless it is clear that the idea of a UN administration received much less support in Argentina than it did in Britain, also that a UN administration compares unfavourably for the Argentine public, with an autonomous Argentine province or joint Argentine-British control.[17] While the results touching on the dispute over

TABLE 12.7 *British and Argentine acceptance of each compromise*
(*Percentage acceptance of each option considered separately, March 1990*)

	UK Total	*Cons*	*Lab*	*Argentine Total*	*PJ*	*Rad*	*DP+ UCD*	*Left*
An Argentine province with guarantees								
Acceptable	33	24	41	87	87	86	100	89
Unacceptable	47	60	41	5	7	4	0	7
Don't know	20	16	18	8	6	10	0	4
Leaseback								
Acceptable	20	11	26	24	23	18	39	28
Unacceptable	56	69	51	62	66	66	61	61
Don't know	24	20	23	14	11	16	0	11
Joint control								
Acceptable	30	22	39	34	30	29	54	28
Unacceptable	53	65	46	56	63	56	46	63
Don't know	17	13	15	10	8	14	0	9
United Nations administration								
Acceptable	57	55	63	14	13	17	15	14
Unacceptable	25	28	22	69	74	72	69	83
Don't know	18	17	16	17	13	11	15	3

the islands produced more party differences in Britain than did the earlier questions, they did not show the Conservatives supporters staunchly maintaining British sovereignty over the islands. Virtually half the Conservatives opted for a United Nations administration as their preferred compromise, while little more than a quarter of the Conservatives rejected it, when it was put directly to them. Once again the results were scarcely compatible with a nationalistic pride generating a continuing 'Falklands factor'.

With this set of restricted questions on the options it becomes very clear that the Argentine government would have no problem selling to its own people a settlement that made the Malvinas an Argentine province, with guarantees to the islanders. On the British side this option was acceptable to a somewhat higher proportion of voters than was leaseback or joint control, but it still clearly failed to obtain majority acceptance. Equally one may note that the Falklands being an Argentine province was acceptable to a third of the British electorate, including a quarter of the Conservative supporters. This result could not be described as demonstrating a widespread national commitment to the islands. Even though we now find the biggest differences between Conservative and the Labour supporters, this question would still not serve to provide clear differentiation between the parties. In March 1990 Labour support was at record levels and this meant the majority of those rejecting the Falklands being an Argentine province were Labour supporters. Any British government would have to work hard to sell such a settlement to the voters and it is difficult to imagine they could do so, unless many of the Falkland Islanders also came to support this solution, a situation that is not foreseeable in the current circumstances.

The overall conclusion on public attitudes to the compromise options is that there is much more willingness to compromise in both countries than is suggested by the narrow official positions on sovereignty. Neither leaseback nor joint control come out well in either country when compared to other options or when considered separately on their own merits. Although leaseback had been the option preferred by the diplomats before the 1982 war, joint control comes out with more support among the general public in the two countries. Nevertheless, it would still appear to have minority support that reduces its viability. The electorates in each country each offer majority support to one of the options, but not to the same one. More research needs to be done to see whether the option preferred by the Argentine electorate of the Falklands becoming a province, with guarantees for the islanders to maintain their way of life, could become more attractive to the British electorate with a tighter specification of the guarantees. Similarly, we do not know whether the option preferred by the British electorate of setting up a UN administration could be made more attractive to the Argentine electorate, by emphasising the removal of the British government presence, even if an Argentine government presence were not established in its place. A further point is that we do not know whether a form of independent, self-governing status – as with the small islands of Kiribati, Tuvalu and Niue – might be attractive on all sides. Finally, of course, the one piece of research nobody has done is to find what leeway for changed attitudes there might be among the islanders themselves, not in the immediate future, but in the medium to long-term.

So far we have considered the Falklands/Malvinas dispute from the point of view of the elites. The governments of Argentina and Britain, the political elites, the relevant pressure groups, business people and others having a direct involvement have been keen to see relations between the two countries improve, but have been very cautious about going ahead. They have also been discussing potential settlements to the dispute in formal terms, but mainly doing so in a private rather than a public debate. It took virtually eight years before the doors were all opened officially for ordinary bilateral relations of all types to be resumed. Yet there is no reason to suppose the two publics felt any need to wait so long. (Attitudes on bilateral relations were no different in Britain in 1990 from attitudes in 1984.) Official positions on sovereignty have barely changed and the 'national interest' is defended unequivocally by both governments. However, the two publics have large minorities willing to accept some form of compromise settlement. The survey results do not suggest that in the 1990s the elites need fear a 'Falklands factor' in either country. Let us now reverse the perspective and ask how the two publics see the situation, in their own terms.

The first question to consider is what ranking the Falklands/Malvinas dispute has, as a current concern, among the many issues competing for the attention of the two publics (see Table 12.8). Early in each of the March 1990 sets of interviews, the respondents were asked 'What would you say is the most urgent problem facing the country at the present time?'. In Britain the mention of the Falklands was zero and in Argentine it was 1 per cent. When the respondents were prompted to name a second problem facing the country and a count was made of the two answers combined, the Falklands still remained at a zero level in Britain and only went up to 2 per cent in Argentina. In both countries domestic economic problems were the prime concern, with inflation followed by unemployment dominating the answers in Argentina and with housing, including the poll tax and mortgages, followed by inflation receiving the most mentions in Britain. (One interesting difference is the environment being given as one of their two answers by 12 per cent in Britain,

TABLE 12.8 *International problems facing Argentina, March 1990*

	First replies given					*First and second replies combined*				
	Total	PJ	Rad	DP+ UCD	Left	Total	PJ	Rad	DP+ UCD	Left
Foreign debt	42	42	44	29	41	64	63	64	72	61
Malvinas	22	21	22	18	24	41	43	41	25	35
Latin American integration	10	10	12	11	22	25	20	29	25	39
Relations with the 'First World'	8	5	8	39	2	15	9	21	60	19
World peace	5	3	4	0	7	15	12	15	7	14
Other	6	4	7	4	2	21	16	21	11	23
Don't know, no reply	7	13	3	0	2	19	33	10	0	9

but 0 per cent in Argentina.) No claim can be made that the Falklands/Malvinas dispute was salient to the two publics in relation to their domestic politics.

The respondents were next asked 'What are the most important international problems facing Britain/Argentina now?'. Again the Falklands received a zero rating among the first responses in Britain, but sufficient 'Falklands' answers did appear after prompting to reach 1 per cent, when the two responses were combined. In Argentina the Malvinas was now mentioned by 22 per cent as their first response and a total of 41 per cent in the two answers combined, but it clearly aroused less mentions than the problem of Argentina's foreign debt. Supporters of the main parties differ little, but supporters of the minor parties are somewhat less likely to mention Malvinas. The right-wing stand out as making relations with the 'First World' as their priority, while the left give somewhat more attention to Latin American integration. In Britain a diverse range of international problems was of concern. The environment was top of the list with 10 per cent among the first answers and 9 per cent among the second. The Common Market, Hong Kong, South Africa, Eastern Europe, world peace and German unity were all mentioned by at least 10 per cent in one of the two answers given by British respondents, whereas from this list only world peace went over 1 per cent in Argentina.

In Britain the situation was unambiguous; the Falklands was

of almost no salience at all in comparison to other contemporary domestic and international problems. In Argentina the official position is that the Malvinas are an occupied section of Argentine territory, yet it clearly does not have this resonance for the Argentine public, because the Malvinas was mentioned as an *international* problem rather than a *domestic* problem (*problema internacional* rather than *problema en nuestro pais*). The significance of the Malvinas as an international problem is not high given that it came second to the foreign-debt problem, and that it was not mentioned at all by 59 per cent of the electorate. In addition, a different question was asked in the August 1990 survey on what should be the government's priority in external relations, with the respondents being offered a closed set of four possible answers. The new question was worded to test for the effect of suggesting improved trade as a priority and the results demonstrate that even in foreign affairs the economy clearly comes first. Obtaining more exports or new markets was chosen by 54 per cent, renegotiating the foreign debt by 17 per cent, promoting Latin American integration by 13 per cent and regaining the Malvinas by 10 per cent. In Argentina too the Falklands/Malvinas dispute has no relevance to the ordinary person's daily life and can be of little salience, unless someone fought in the 1982 war or had a relative or friend killed in that war. The dispute may be able to generate some emotion in an abstract debate, but its low salience to the average member of the British public or the Argentine public means it is unlikely to affect voting patterns.

We hypothesise that the two publics see the issue as involving four dimensions. There is a symbolic conflict, a concrete dispute, the problem of the future of the islands and the problem of the future of the islanders. Firstly, when people debate 'sovereignty' they are invoking a highly abstract term that is ill-defined and understood in many different ways, even among the elites, let alone in the general publics. It is a symbolic assertion of 'national identity'.[19] Secondly, the war and its aftermath (particularly for the bereaved), current military and diplomatic positions, the debates in the UN and the implications of the dispute for other practical problems, such as fishing around the islands, are all immediate concrete concerns for government policy. Most of the time policy is handled by junior officials, but when a new decision has to be taken it may come to the attention of the two publics. Thirdly, at some point in the future, there may be an attempt to reach a settlement about the territory that is acceptable on all sides. Finally, any future settlement will have to define the rights of the islanders. At the elite level these four dimensions are seen as being intimately related. They are all part of a single dispute that is normally summarised as being a sovereignty dispute. For the lawyers and the politicians, it is difficult to see any aspect of Argentine-British relations or of the Falklands/Malvinas question distinctly from the overall problem of sovereignty. We believe that the publics do not perceive the dispute as a single coherent issue, with everything constrained by 'sovereignty'; they do see the four dimensions separately.

That *Las Malvinas son Argentinas* is an Argentine statement of faith, which will rarely elicit a contrary opinion when voiced to a fellow Argentine. Similarly in Britain 'The Falklands are British' may taken for granted as a simple pragmatic statement of fact. For a few weeks in 1982 these positions were reversed when in Argentina it became a matter of fact and in Britain a matter of faith. For Argentines the Malvinas are part of the territory of a large and important country that extends from the tropics to the South Pole and from the Andes mountains through the Patagonian pampas to dominate the South Atlantic. There is a significance attached to territory that is virtually absent from British culture. Prior to 1982 very few British people would have been aware of the existence of the Falklands as a British possession. Since 1982 there has been a pride in the achievements of the Task Force. For some in Britain it may be a reawakening of imperial glories, for others a commitment to the principles of resisting aggression and defending democracy, and for others a triumph of pragmatic muddling through, in which rapid improvisation produced a response of a scale and a speed that few would have thought possible before it actually happened. Much as these emotions of attachment to Argentine territory or pride in British achievement are a part of the respective national identities, they are not central to current politics. We have seen that the Falklands/Malvinas dispute is no longer salient to the voters. While it might be argued that the British public are not concerned with the dispute because they can be content with the *status quo*, this is not the case for the Argentine public. Nevertheless, the islands have routinely been mentioned by only 1 or 2 per cent as the number one problem. This lack of priority has been recorded in over twenty SOCMERC polls from 1984 to 1990. The emotions of national identity that the symbolic question of sovereignty invokes will not guide policy; they set constraints on what is possible, boundaries for political debate.

We can see the separation between the symbolic aspects of sovereignty and day-to-day questions of policy in a further question that was asked about contacts between politicians from Britain and Argentina (see Table 12.9). This question was deliberately biased by invoking an emotional argument against having contacts. The results show that the suggestion that 'it is disloyal and disrespectful to those who died in the war, if we have contacts' with politicians from the other country, is rejected by a substantial margin in each category among the electorates

TABLE 12.9 *Reactions to charges of disloyalty*

Some people say that it is disloyal and disrespectful to those who died in the war, if we have contacts with Argentine politicians. Do you strongly agree, agree, disagree or strongly disagree with this?
(UK results, March 1990)

				Lib	*Occupational Class*		
	Total	*Cons*	*Lab*	*Dem*	*A/B/C1*	*C2*	*D/E*
Strongly agree	5	11	4	2	5	7	5
Agree	17	15	17	16	12	17	22
Disagree	52	57	50	48	55	53	48
Strongly disagree	15	9	18	24	18	11	14
Don't know	11	8	11	9	10	11	11

Some people say that it is disloyal and disrespectful to those who died in the war, if we have contacts with English politicians. Do you strongly agree, agree, disagree or strongly disagree with this?
(Argentine results, March 1990)

				DP+		*Upper/*		*Lower*	
	Total	*PJ*	*Rad*	*UCD*	*Left*	*U Mid*	*Middle*	*Middle*	*Lower*
Strongly agree	5	4	4	4	9	5	3	4	8
Agree	20	20	21	14	13	14	24	21	23
Disagree	56	51	62	71	63	64	50	54	47
Strongly disagree	7	11	4	11	0	8	11	4	3
Don't know	13	13	9	0	15	9	12	16	19

of the two countries. A slight party difference was found on this question in Britain in February 1986, but in the opposite direction to that expected: Conservative supporters were more likely to disagree with the charge of disloyalty. Again in Argentina the supporters of the main parties, but not the minor parties, are remarkably close to those in Britain, with the right-wing being noticably more conciliatory towards the British and the left being somewhat less conciliatory. In both countries there is a definite expression of class differences in the responses, with slight differences based on sex and age: the working class, women and older people being the less likely to reject the charge of disloyalty. Overall there are two striking aspects to these results. Firstly, there is no significant difference between the two countries; secondly, deliberate invocation of the symbolic dimension, in an emotional form, does not result in opposition to improving bilateral relations.

SOCMERC has investigated basic Argentine attitudes to government policy several times, by asking 'What do you think Argentina should do concerning the Malvinas?' (see Table 12.10) Initially in April 1982 the two dimensions of symbolic sover-

TABLE 12.10 *Argentine policy towards the dispute*

What do you think Argentina should do concerning the Malvinas?

	Dec 1984	*June 1985*	*Aug 1985*	*Oct 1985*	*Apr 1986*	*March 1990*
Negotiate	37	43	47	52	56	58
Demand the Islands' return	46	40	41	35	31	30
Pursue the conflict	5	6	5	4	4	1
Don't know, no reply	12	12	7	8	10	10

	March 1990 sample Total	*PJ*	*Rad*	*DP+ UCD*	*Left*	*Upper/ U Mid*	*Middle*	*Lower Middle*	*Lower*
Negotiate without conditions	19	14	23	46	4	24	18	16	1
Negotiate about sovereignty	39	35	39	43	50	44	42	38	22
Demand the Islands' return	30	37	32	11	43	25	28	34	42
Pursue the conflict	1	1	1	0	0	1	3	2	1
Don't know, no reply	10	11	5	0	2	6	9	10	25

eignty and of conflict policy were confused with each other. For some it was a question of the ends justifying the means and for others it was not apparent that war was imminent. As soon as the fighting was over a clear trend set in. In late 1982 30 per cent of the electorate thought that Argentina should somehow continue to 'pursue the conflict' ('*proseguir el conflicto*', which carries active connotations), but by June 1984 this was down to 7 per cent. In these questions the respondents were expected to choose simply between 'pursue the conflict' and 'negotiate'. As the total of 83 per cent in favour of negotiating was by June 1984 so very high, in subsequent surveys an intermediate option of 'demand the return of the Islands' was included. In April 1986 the negotiate option was split into 'negotiate without prior conditions' and 'negotiate only if the subject of sovereignty is included'. The results show stability, but with a trend across time: more people are coming to support negotiations and there is a steady decline in the relatively hard-line desire to demand the return of the islands. Even the small group wanting to pursue the conflict actively seems to be dwindling away. The breakdown by party preference in the most recent poll does not show any major differences between the *Justicialistas* and the Radicals. However, this question produces the sharpest differentiation in the whole survey between the positions taken by the supporters of the minor parties, with a much more conciliatory set of responses from the right wing and the left taking the hardest stand.

The interviewing in March 1990 was carried out only a short while after Britain and Argentina had concluded a successful round of negotiations on bilateral relations. It is interesting that this did not influence the poll's results and lead to a general increase in support for negotiations on the Malvinas. Furthermore, the 1990 Madrid negotiations with sovereignty being 'under an umbrella' has been seen as very much the product of President Menem personally, yet the *Justicialistas* gave less support to negotiations on the Malvinas than did the Radicals. While the policy positions chosen did not vary by sex or age, the table shows that there were substantial variations by class. Both the negotiation positions received only half as much support among the lower class as among the upper class and support for negotiation gave way to the desire to demand the return of the islands. As party choice is related to class in

Argentina, the party differences on Malvinas policy may well be explained by these class-based differences in attitudes rather than being a product of party loyalties.

The equivalent question asked by Gallup in Britain covering the government's tactics was 'What do you think that Britain should do now about the Falkland Islands?' (see Table 12.11). The options given were one hard-line position of 'nothing except maintain the base' and three alternative approaches to negotiations. The hard-line position receives a similar level of support to the equivalent position in Argentina and does not vary between the two main parties. However, party differences do appear on the question of whether sovereignty can be discussed. Just under two-thirds of both Conservative and Labour supporters opted for some form of talks with the Argentines, but twice as many Conservatives as Labour supporters wanted the qualification that sovereignty must be excluded. Nearly a third of the Conservatives were willing to go against the British government's position by including sovereignty, while nearly a half of the Labour supporters were willing to do so.

On average the British population is almost equally divided into four groups, one taking a harder line than the government, one endorsing the government's position and two being

TABLE 12.11 *British policy towards the dispute, March 1990*

What do you think Britain should do now about the Falkland Islands?

	UK Total	*Cons*	*Lab*	*Men*	*Women*	*Occupational class* A/B/C1	C2	D/E
Nothing, except maintain the base	24	28	26	22	27	18	27	29
Talk with Argentina on all subjects except sovereignty	24	33	15	30	18	28	24	18
Have talks allowing the Argentines to state their case on sovereignty	20	16	23	18	21	21	20	18
Negotiate with Argentina about sovereignty	20	13	25	23	17	23	18	18
Don't know	12	11	12	7	17	10	10	16

more conciliatory than the government. A slight sex difference shows women tending to be a little more hard-line than men. Yet another result for Britain is similar to that in Argentina: there was an effect of class upon attitudes, though not as marked as in Argentina. The A/B/C1 groups of professional, managerial and skilled working-class people were less willing to do nothing and more willing to negotiate, the main difference being in the proportion wanting to talk excluding sovereignty, with only a small increase in those from this upper- and middle-class group being conciliatory enough to include sovereignty.

Thus in both countries there are clear majorities in favour of negotiations. Only a minority in each country translate the symbolic commitment to sovereignty into a hard-line government policy on sovereignty. In neither country are there bellicose attitudes. This is shown most dramatically in Argentina by the drop in support for 'pursuing the conflict', from 30 per cent in late 1982 to 1 per cent in early 1990. It is also shown in Britain by only a quarter endorsing the 'Fortress Falklands' approach of doing nothing except maintaining the base. The earlier analysis of questions on bilateral relations shows no disposition of the two publics to regard each other as enemies.

In the discussion of attitudes to various proposals for settlement of the dispute, we found that there was far from unanimous commitment to control over the territory. Only a third of the British respondents said the Falklands should be permanently British and only two-thirds of the Argentine respondents said the Malvinas should be handed to Argentina. In addition to large minorities being willing to accept some form of compromise, a tenth of the British were willing to abandon their claim. The symbolic commitment to sovereignty did not necessarily translate into a commitment to the territory. However, sovereignty is also about the people of a territory or, put in more pragmatic terms, another dimension of the issue is the future of the inhabitants of the Falklands.

In Argentina in April 1986 SOCMERC asked their sample a question about attitudes to the people on the islands, the 'Kelpers' as they are called colloquially. This question was then repeated in the March 1990 survey in both countries (see Table 12.12). In 1986 as many as a quarter of the Argentinians were unable to give an opinion about the islanders and less than half were willing to concede any right of the islanders to affect the

TABLE 12.12 *The rights of the Falkland Islanders*

What do you think about the Islanders?

	March 1990 UK Total	*Cons*	*Lab*	*March 1990 Argentine Total*	*PJ*	*Rad*	*DP+ UCD*	*Left*	*April 1986 Argentine Total*
They have the right to decide about the future of the Islands	63	70	60	24	26	21	21	22	15
Their interests should be taken into account, but they should't be the ones to decide	26	21	27	37	32	39	43	46	27
They are a secondary factor and they must adapt to whatever is agreed for them by Britain and Argentina	4	2	6	29	30	34	32	24	34
Other answer given	1	2	1	2	0	2	4	9	–
Don't know	6	5	6	8	11	5	0	0	25

situation. By 1990 attitudes had changed substantially in Argentina in two ways: the 'don't know' response had dropped down to the more usual level for most attitude questions (below 10 per cent) and recognition of the Islanders' rights had risen to 61 per cent.

Two answers offered to the respondents recognise the islanders' rights. The stronger version, 'they have the right to decide about the future of the Islands' corresponds quite closely to the view being taken by Mrs Thatcher, then the British Prime Minister, that their wishes must be 'paramount'. The weaker version, that 'their interests should be taken into account, but they shouldn't be the ones to decide', corresponds to the position in Britain of the Labour Party, the Liberal Democrats and many Conservative backbench MPs. It is also the position adopted by most of the political elite in Argentina. Thus it is somewhat surprising that a sizeable proportion of the Argentine electorate – a quarter – chose the stronger version of the islanders' rights. These responses did vary a little by party and somewhat more by class, with the lower class being less willing to recognise the islanders' rights. Support for the stronger version of the islanders' rights was relatively constant and the variation was mainly in the level of support for the weaker version of their rights.

The lack of a more marked variation by party is contrary to expectations in the case of the right-wing parties. While the right wing did show themselves, on the previous question, to be much more sympathetic to the British desire to negotiate without conditions, their British links and pro-British attitudes did not translate into greater awareness of the islanders and their importance in British politics. On the other hand, the left, who took a much harder line about negotiations, had more sympathy for the islanders than any of the other parties. As the main targets of the military repression in the 1970s, the hard left has mainly been distinguished since then by their punitive stance towards those who violated human rights. Thus, the fact that this particular question was evoking reactions both to the Falklands dispute and to human rights explains the responses from the minor parties not fitting into their usual patterns.

So far we have found similarity between Argentine and British responses on many questions. The two electorates were very close in supporting improvements in bilateral relations; they clearly differed on a crude summary question about the future of the islands, but they were closer in their acceptance of various possible compromise solutions; and both sides were conciliatory in wanting some form of negotiating process. When we consider the British response to the question on the rights of the islanders, we find for the first time in these surveys that a clear majority of the British public took a position contrary to that of both the Argentine government and the Argentine public. Support for the rights of the islanders does not drop below 85 per cent for any category of party preference, sex, age or class, and it is the stronger version rather than the weaker version of their rights that is chosen by at least a two-to-one margin in all categories. Much of the political elite in Britain would refuse to make the wishes of the islanders paramount, but 63 per cent of the electorate endorsed this position. There is some party variation around this average figure, with 70 per cent of the Conservatives and 60 per cent of the Labour supporters going for the stronger version of the islanders' rights, but this hardly constitutes a significant division between the parties. The result is all the more striking in comparison to the relative lack of sensitivity about sovereignty in the previous questions. For the electorate the British commitment is to the people of the Falklands and not to sovereignty over the territory. While only a third wanted the islands to be permanently British and a majority were willing to hand over to the United Nations, nearly two-thirds thought the islanders 'have the right to decide about the future of the Islands'.

We conclude that both governments can feel pleased that the agreements reached in Madrid in February 1990 are warmly welcomed by the peoples of the two countries. Major differences remain about the long-term future of the Falkland Islands, but both sides are willing to consider some form of compromise settlement. Debate is needed in Argentina on how far they are willing to go in making guarantees to the islanders that their way of life could be maintained. If the Argentine government could convince the islanders that guarantees of their rights were permanent and reliable (and for the time being that is a very big 'if'), then a long-term agreement would clearly be possible and acceptable to the voters of both Britain and Argentina.

NOTES

1. The South Atlantic Council is grateful to the Centre for Conflict Analysis and Resolution at George Mason University for making funding available from a grant by the Glad Foundation, to enable it to commission the March 1990 survey in Argentina. Without this support the research could not have been carried out.
2. For more details of the historical background, see Peter Beck, *The Falkland Islands as an International Problem* (London: Routledge, 1988). The main discussions of the various options for the future of the Falklands are in House of Commons Foreign Affairs Committee, *Falkland Islands, Vol. I, Report and Minutes of Proceedings*, (London: HMSO, HC(1983–84) 268–I, 25 October 1984), pp. xlii–xliv; *Minutes of Evidence*, (London: HMSO, HC(1982–83) 31–i to 31–xiv, November 1982 to February 1983); and *Minutes of Proceedings* (London: HMSO, HC (1982–83) 380, 11 May 1983), pp. xlvi–lv. W. Little and C.R. Mitchell,

In the Aftermath: Anglo-Argentine Relations since the War for the Falklands/ Malvinas Islands (Maryland: Center for International Development and Conflict Management, University of Maryland, 1989); B. George MP and W. Little, *Options in the Falklands-Malvinas Dispute*, (London: South Atlantic Council, Occasional Paper No. 1, April 1985, out of print; reprinted in Little and Mitchell, pp.141–52); and P. Beck, op. cit., Chapter 6.

3. Information from Peter Willetts's conversations with various diplomats and former diplomats.
4. J.R. Vanossi, 'Provincialisation with a Pact of Incorporation' (circulated at a University of Maryland conference and reprinted in Little and Mitchell, pp. 111–20).
5. D. Bullock and C.R. Mitchell, *The Aland Islands Solution* (London: South Atlantic Council, Occasional Paper No. 3, March 1987; reprinted in Little and Mitchell, pp. 153–59).
6. The Rt Hon. David Owen MP put the case for the Falklands to become a strategic UN Trust Territory under the Security Council in a debate on 'The Falkland Islands (Foreign Affairs Committee Report)' on 14 March 1985. See *Hansard Official Report, Sixth Series, Commons Vol. 75 1984–85*, c508–9.
7. M. Dent, *Shared Sovereignty: A Solution for the Falklands/Malvinas Dispute*, (London: South Atlantic Council, Occasional Paper No. 5, March 1989). An earlier version is in Little and Mitchell, pp. 121–40.
8. The scheme has been outlined in a paper circulated privately by Mrs Edna Lemle.
9. Within the Island community there is deep suspicion of the Foreign Office and some hostility towards the expatriate civil servants; comments by Councillors to the UN Special Committee on Colonialism have sometimes suggested that they sought greater autonomy from Britain; but the Marplan poll released by the UK Falkland Islands Committee on 26 March 1986 showed 94.5 per cent of the Islanders wanted to remain British, with only 1.6 per cent opting for independence.
10. Social Surveys (Gallup Poll) Ltd carries out weekly 'omnibus' surveys with the respondents coming from national quota samples controlled for sex, class and region. Usually there are approximately 1,000 respondents. Despatch of the main survey in March 1990 was slightly affected by a local postal strike, with the result that the sample was only of 856 voters, interviewed from 28 February to 5 March 1990. In Argentina 800 adults were interviewed by SOCMERC (Mora y Araujo, Noguera and Associates) in the main survey from 1–13 March 1990, from a random national sample controlled for sex and region.
11. The answers given to these questions in each survey were not strictly comparable, partly because the passage of time put them each in a different general political context and partly because the questionnaires were slightly different from each other. In particular the results for 1984 and 1986 cover steps that had yet to be agreed between the two governments, whereas the results for 1990 cover steps that had by then been agreed at the February 1990 Madrid talks. More details are given in Peter Willetts and Felipe Noguera, *Prospects for a Settlement of the Falk-*

lands/Malvinas Dispute: An Analysis of Public Opinion in Britain and Argentina (Center for Conflict Analysis and Resolution, George Mason University, 1990).

12. The debate about the effect of the Falklands war upon public opinion has produced an extensive academic literature. See particularly: I. Crewe, 'How to Win a Landslide Without Really Trying: Why the Conservatives Won in 1983', pp. 155–96 of A. Ranney (ed.), *Britain at the Polls* (New York: Duke University Press, 1985); P. Dunleavy and C. T. Husbands, *British Democracy at the Crossroads: Voting and Party Competition in the 1980s* (London: George Allen and Unwin, 1985); H.D. Clarke, M.C. Stewart and G. Zuk, 'Politics, Economics and Party Popularity in Britain 1979–83', *Electoral Studies*, 5 (1986) pp. 123–41; D. Sanders, H. Ward, and D. Marsh, with T. Fletcher, 'Government Popularity and the Falklands War: A Reassessment', *British Journal of Political Science*, 17 (1987) pp. 281–313; H. Norpoth, 'The Falklands War and Government Popularity in Britain: rally without consequence or surge without decline?', *Electoral Studies*, 6 (1987) pp. 3–16; H. Norpoth, 'Guns and Butter and Government Popularity in Britain', *American Political Science Review*, 81 (1987) pp. 949–959; H. Norpoth, 'The Falklands Factor. The Latest Blast', *Contemporary Record*, Winter 1988, p. 26; H.D. Clarke, W. Mishler, P. Whiteley, 'Recapturing the Falklands: Models of Conservative Popularity, 1979–1983', *British Journal of Political Science*, 20 (1990) pp. 63–81; and D. Sanders, H. Ward and D. Marsh, 'A Reply to Clarke, Mishler and Whiteley', ibid., 20 (1990) pp. 83–90.
13. The eighteen questions on Mrs Thatcher's image in Gallup's July 1984 survey, when correlated with the four questions on Argentine–British relations, gave seventy-two correlation statistics that might indicate some evidence of a 'Falklands factor'. The highest correlation turned out to be 0.13 and the seventy-one others were less than 0.1, so no relationships existed between these variables.
14. The *Daily Mail* report of 4 October 1986 suggested that British soldiers objected to the sympathetic treatment given to Sñr Isaias Gimenez and his daughter Maria, mocked the Foreign Office official who met them and said some islanders were 'shocked'.
15. The *Daily Telegraph* of 4 October 1986 said 'Mr Robin Pitaluga a former member of the legislative (*sic*) attacked the presence of Argentine relatives at the funeral' of Lt Gimenez. *The Times* of 31 March 1990 quoted Tricia Card saying of Argentines in general 'We don't want anything to do with them. We don't even want them to come and visit the cemetery, . . . although I suppose that's all right as long as they don't stay'.
16. The British government has officially accepted that visits by Argentine relatives to the war graves on the islands may be organised by the Red Cross (Prime Minister's answer to a Written Question, 31 March 1983 and many statements since then). However, it has not actively pursued the question, to ensure that the visits do take place. The Argentine government did not publicly endorse the possibility of such visits until the communiqué from the Madrid talks in February 1990.
17. The poll in Argentina in August 1990 was less detailed than the one in March 1990. Four changes in the questionnaire biassed the August

1990 results towards a lower Argentine level of acceptance of controversial compromise options than was shown in the March 1990 survey. In August 1990 a UN administration was accepted by marginally more Argentine voters than was leaseback. It is a reasonable assumption that this would also have occurred if the question had been asked in March 1990. On this basis the figures for Argentina in Table 12.7 might well have been 26–30 per cent acceptance and 54–58 per cent rejection of a UN administration.

18. We do not have space here to discuss challenges that might be made to common assumptions about the nature of any supposed Argentine or British national identity.

Index